Online Identity Theft

ORGANISATION FOR ECONOMIC CO-OPERATION AND DEVELOPMENT

The OECD is a unique forum where the governments of 30 democracies work together to address the economic, social and environmental challenges of globalisation. The OECD is also at the forefront of efforts to understand and to help governments respond to new developments and concerns, such as corporate governance, the information economy and the challenges of an ageing population. The Organisation provides a setting where governments can compare policy experiences, seek answers to common problems, identify good practice and work to co-ordinate domestic and international policies.

The OECD member countries are: Australia, Austria, Belgium, Canada, the Czech Republic, Denmark, Finland, France, Germany, Greece, Hungary, Iceland, Ireland, Italy, Japan, Korea, Luxembourg, Mexico, the Netherlands, New Zealand, Norway, Poland, Portugal, the Slovak Republic, Spain, Sweden, Switzerland, Turkey, the United Kingdom and the United States. The Commission of the European Communities takes part in the work of the OECD.

OECD Publishing disseminates widely the results of the Organisation's statistics gathering and research on economic, social and environmental issues, as well as the conventions, guidelines and standards agreed by its members.

This work is published on the responsibility of the Secretary-General of the OECD. The opinions expressed and arguments employed herein do not necessarily reflect the official views of the Organisation or of the governments of its member countries.

Foreword

OECD member countries understand the need to better address the challenges arising from misuse of the Internet. The "Future of the Internet Economy" project was initially presented to the OECD Committee on Consumer Policy (CCP) at its October 2005 and March 2006 meetings, with an invitation to the countries to participate. Recognising the close synergies with its past and future work programme in relation to this issue, the CCP, at its 72nd session in October 2006, agreed to provide input to the project. Cyber fraud was identified as the main theme to be examined, with a particular focus on ID theft; mobile commerce; user education and empowerment; cross-border law enforcement co-operation; and redress for individual victims on line.

Unlike most of the subject areas identified above, ID theft is a new topic for the CCP to explore. This issue is indirectly covered by the *1999 OECD Guidelines for Consumer Protection in the Context of Electronic Commerce* ("The 1999 *E-commerce Guidelines*") (OECD, 1999), and the *2003 OECD Guidelines for Protecting Consumers from Fraudulent and Deceptive Commercial Practices Across Borders* ("The *2003 Cross-border Fraud Guidelines*") (OECD, 2003). But ID theft has never been, as such, the subject of an in-depth study by the Committee. In light of the growing importance of this issue, at its 72nd session, the CCP requested the Secretariat to prepare an analysis of online ID theft, to better understand the concept and the extent to which it can affect consumers and users.

ID theft is a cross-cutting problem that violates not only consumer protection rules, but also security, privacy, and anti-spam rules. Considering how best ID theft could be countered should be done under a multi-disciplinary approach. Thus, in preparing the initial scoping paper, the CCP Secretariat consulted with the Secretariat to the OECD Working Party on Information Security and Privacy (WPISP). The CCP Secretariat also established contacts with the OECD's Working Party on Indicators for the Information Society (WPIIS) to examine how statistics could better measure online ID theft cases.

The resulting *Scoping Paper on Online Identity Theft* was discussed by the Committee on Consumer Policy ("CCP") at its 73rd and 74th Sessions in 2007. It was prepared by Brigitte Acoca, of the OECD Secretariat, based on independent research and member countries' input. It served as a basis for the development, in 2007 and 2008, of the CCP's policy principles on online identity theft. The paper was declassified by the Committee on Consumer Policy on 9 January 2008, under the written procedure.

The following book is based on The Scoping Paper on Online Identify Theft, and the CCP's policy principles on online ID theft: OECD Policy Guidance on Online Identify Theft.

Table of Contents

Executive Summary

Identity theft is an old misdeed, but the growth of Internet and e-commerce has taken it to a whole new level. Using widely available Internet tools, thieves go "phishing" and "pharming" to trick unsuspecting computer users into providing personal data, which they then use for illicit purposes. All of this is possible because face-to-face client relationships do not exist on the Internet. Establishing one's real identity for online transactions is complicated, thereby making fraud easier.

The potential for fraud is a major hurdle in the evolution and growth of online commerce. E-payment and e-banking services – the focus of this book – suffer substantially from public mistrust. In the United Kingdom, for example, a recent study estimated that 3.4 million people were unwilling to shop online owing to concerns about online security.

Given the growth of online ID theft, many OECD member countries have taken steps to ensure that consumers and Internet users are adequately protected. These steps encompass various measures: consumer and user-awareness campaigns, new legislative frameworks, private-public partnerships, and industry-led initiatives focused on technical responses.

Despite these initiatives, most countries have not sufficiently addressed the problem of ID theft, and online ID theft in particular. Thus, the scope, magnitude and impact of ID theft vary from one country to another, or even from one source to another within the same country. These differences reflect the need for a more co-ordinated response at both the domestic and international levels.

The purpose of this book is threefold: (1) to define ID theft, both online and off-line, and to study how it is perpetrated, (2) to outline what is being done to combat the major types of ID theft, and (3) to recommend specific ways that ID theft can be addressed in an effective, global manner.

The tools of the trade

As explained in Part I of this book, there are three main methods thieves use to obtain personal information via the Internet and/or individual computers:

- Malicious software (malware) is surreptitiously installed into a computer or device – fixed or mobile – to collect the user's personal information over time.
- Computers or mobile devices are hacked into, or otherwise exploited, to obtain the user's personal data.
- "Phishing," whereby thieves use deceptive e-mails to get users to divulge personal information, includes luring them to fake bank and credit-cards websites. These phishing messages, commonly distributed by e-mail spam, are also used to install malware on the computers of unsuspecting recipients.

Phishing techniques are becoming more sophisticated and harder to detect. Some aptly named principal forms are:

- "Pharming": using deceptive e-mail messages to redirect users from an authentic website to a fraudulent one, which replicates the original in appearance.
- "SMiShing": sending text messages ("SMS") to cell phone users that trick them into going to a website operated by the thieves. Messages typically say that unless users go to the website and cancel, they will be charged for services they never actually ordered.
- "Spear-phishing": impersonating a company employee/employer via e-mail in order to steal colleagues' passwords/usernames and gain access the company's computer system.

ID thieves misuse victims' personal information for a plethora of unlawful schemes. Typically, these involve: misuse of existing accounts; opening new accounts; fraudulently obtaining government benefits, services, or documents; health care fraud; and the unauthorised brokering of personal data.

Challenges in addressing online ID theft

Part II of this book outlines what is currently being done to address online ID theft at the domestic and international levels. There are ongoing efforts, but they are stymied by two major issues:

1. *Lack of a common definition.* ID theft is defined differently in OECD member countries: some view it as a specific crime, while others regard it as a preparatory step in the commission of other wrongs or crimes. Lack of a common definition for ID theft may complicate efforts to combat the problem in a comprehensive, cross-border fashion.

2. *Lack of comparable data.* ID theft (whether offline or online) has largely failed to attract the attention of statisticians. Most data are from the United States; statistics for Europe are sparse, except for the United Kingdom. When data are available, they often do not cover ID theft as an independent wrongdoing. The United States is one of the few countries with data that analyse ID theft as a separate offense.

In addition, statistics are collected differently by countries, complicating cross-border comparisons. Moreover, findings of public and private entities vary greatly: some sources conclude that the scale of ID theft has declined in recent years, resulting in growing consumer confidence. Other sources show ID theft is on the rise.

Prevention and enforcement strategies

Domestic level: As discussed in Part II, the public and private sectors in most OECD member countries rely on a range of consumer and user-education tools to combat ID theft. In some countries, information is shared among various public and private entities in order to investigate and prosecute ID theft. Resources for policing are limited, however, in many countries.

Companies in some countries understand the need to deploy human resources and security strategies to prevent data leakage. Nevertheless, there is a recognition that more needs to be done to ensure they adequately prevent those data security breaches.

A few countries have taken steps in this direction, imposing an obligation on data collectors, such as requiring companies and Internet Service Providers (ISPs) to disclose data breaches affecting customers and the public. Other countries are still considering whether such disclosure should be mandated by law. In some European Union member states, ISPs are requesting the right to act on behalf of their customers if personal information is misused and results in direct or indirect losses.

International level: Various international organisations and groups are involved in fighting cyber fraud. The OECD, for example, has developed policy responses in the areas of fraud against consumers, spam, security and privacy, including the *2006 Anti-Spam Toolkit* and the *2003 Cross-border Fraud Guidelines*. In addition, there have been a number of bilateral, multilateral and regional initiatives in which law enforcement, police and government entities have joined forces to prevent and prosecute online ID theft.

Public-private international initiatives: Some efforts are data-sharing fora for gathering statistics on phishing, malware and other online threats. Other efforts are enforcement-oriented, in which the private sector assists governments with investigations, implementing technology, and developing customised legislation and best practices for stemming online ID theft.

Conclusions and recommendations

The analysis in this book suggests that policy makers should address a number of issues in order to advance the fight against ID theft:

- ***Definition*** – The lack of a common definition of what constitutes identity theft may stymie efforts to address the problem in a comprehensive fashion, across borders.
- ***Legal status*** – ID theft/fraud is not an offence per se under most OECD member country laws. It is a crime in a few. Whether ID theft should be treated as a stand-alone offence, and criminalised, needs to be considered.
- ***Co-operation with the private sector*** – The private sector should actively participate in fighting ID theft. OECD member countries could consider more restrictive laws that increase the penalties imposed on ID thieves and could engage in outreach to the private sector to encourage entities to: i) launch awareness campaigns; ii) develop industry best practices; and iii) develop and implement technological solutions to reduce the incidence of ID theft.
- ***Standards*** – Member countries should examine establishing national standards for private sector data-protection requirements. They should consider requiring the disclosure of data-security breaches at companies and other organisations that store data about their customers.
- ***Statistics*** – The production of more tailored and accurate statistical data, covering all OECD member countries, could help determine the impact of ID theft on the digital marketplace.

- ***Victim assistance*** – OECD member countries could consider developing assistance programs to help victims of ID theft recover/minimise their injury.
- ***Remedies*** – Member countries could consider whether to enact legislation to provide more effective legal remedies for victims of ID theft.
- ***Deterrence and enforcement*** – The lack of criminal laws prohibiting ID theft and the limited resources of law enforcement authorities indicate there is insufficient deterrence. Member countries could explore the value of increasing resources for law enforcement, ID theft investigations and training. More generally, given the rapid evolution of ID theft techniques and methods, more resources and training could be granted to all OECD authorities involved in the battle.
- ***Education*** – Consideration could be given to broadening education on ID theft so as to cover all interested stakeholders including consumers, end users, governments, businesses and industry.
- ***Co-ordination and co-operation*** – Agencies involved in the enforcement of ID theft rules and practices are numerous at both domestic and international levels. Their respective roles and framework for co-operation could be clarified to help enhance their effectiveness.

Consideration could be given to improving domestic law enforcement co-ordination by developing national centres dedicated to the investigation of ID theft crimes. With regard to co-ordination and co-operation with foreign law enforcement authorities, member countries could explore areas of mutual interest, such as: i) enhancing deterrence; ii) expanding participation in key international instruments (e.g. the Council of Europe Convention on Cybercrime); iii) and improving response to requests for investigative assistance; and iv) otherwise strengthening co-operation with foreign partners (e.g. in training law enforcement).

This book addresses issues and concepts of a technical nature which might evolve rapidly. The laws may have changed since first publication, and the reader is cautioned accordingly.

Part I. The Scope of Online Identity Theft

Part I of this book examines the multiple facets of the ID theft problem and the different methods used to perpetrate ID theft via the Internet. Chapter 2 describes the techniques implemented by ID thieves to lure victims online, which increasingly involve the use of spam and malware, and how thieves abuse this stolen information. Chapter 3 focuses on ID theft victims. It draws on the available statistical data reflecting victims' complaints and losses trends, and it questions whether ID theft is more prevalent off line than on line.

Chapter 1. The Problem Posed by Online Identity Theft

A new Internet landscape

Over the past 10 years, the Internet has evolved into a single and integrated infrastructure where audiovisual media, publishing, and telecommunications are all converging. This low-cost and seamless communications system not only fosters growth for existing and new industries but also serves society by promoting diffusion of culture and knowledge. Today, the Internet is enhancing commercial opportunities for businesses. It is also serving as a vehicle for providing public services directly to businesses and consumers, as well as innovative personal and social activities. As such, the Internet has substantially changed both our global economy and society and its impact in the coming years is expected to be significant.

Anticipating this evolution, as early as in 1998,[1] the OECD pointed out the importance of sustainable electronic transactions for the global economy and society. At the same time, however, it warned its member economies about the dangerous sides this changing scheme could bring, one of them being the emergence of new types of online threats to the detriment of consumers and users. By nature, face-to-face client-relationships are nonexistent on the Internet. Establishing one's real identity for online transactions is complicated, thereby making fraud easier.

What is identity theft?

There is no standard definition of online, or offline, ID theft at the international level. While some countries have adopted a broad view of the concept, which usually applies to both on and offline ID theft, very few consider it as a specific offence. As a result of these different approaches, the legal nature of ID theft varies from one jurisdiction to another, leading to the application of different regimes of prevention, prosecution and sanctions.

ID theft is an illicit activity with multiple facets. It is generally included in a larger chain of wrongs or crimes. More specifically, ID theft is committed over different sequences of actions. This complexity has opened the path for different legal categorisations of the concept in OECD member countries, which either qualify ID theft as a specific crime, a civil wrong, or as a preparatory step in the commission of other offences such as fraud, forgery, terrorism, or money laundering.

In the absence of a globally accepted definition, this book will use the term "ID theft" as follows:

> *ID theft occurs when a party acquires, transfers, possesses, or uses personal information of a natural or legal person in an unauthorised manner, with the intent to commit, or in connection with, fraud or other crimes.*

Although this definition encompasses both individuals and legal entities, focus in the present paper is limited to identity theft affecting consumers.

While some OECD member countries employ different terminology to describe the problem,[2] all of the countries addressing the issue aim to prevent fraudulent or criminal activity resulting from the misuse of personal information.[3]

ID theft's main elements

The concept of "identity" and "personal information"

ID theft is a problem involving personal information. Our society is increasingly relying on personal information to identify individuals in many circumstances. For instance, it is used to establish accounts with merchants, ISPs, phone companies, *etc.* It may also be used to get access to various accounts and record systems with financial institutions, health organisations, schools, government agencies, and other entities.

Understanding the concept of "identity" and how its components operate in different media is crucial to determine the appropriate means to protect it. The core components of identity are relatively easy to grasp. They are generally based on fixed and verifiable attributes, which are usually officially provided and registered by public authorities.[4] These attributes include individuals' gender, first and last name, date and place of birth, parents' first and last name and in some countries, individuals' assigned social security number.[5] Individuals also can be identified with a variety of other attributes including a computer username and password, a web page, a blog, an Internet Protocol ("IP") address that identifies computers on the Internet, an e-mail address, a bank account and PIN number.

An effective fight against online ID theft will be difficult if the ways in which the different components of identity are used in the online medium are not defined. As concluded in a 2006 *Security Report on Online ID theft*, "after all, we cannot protect what we cannot define," (BT, 2006). Responding to this issue is thus a first essential step to determine how digital personal information can be made harder to steal. In Korea, an improved online identity system was introduced in October 2006 to help verify Korean citizens' identity in cyberspace.[6] The old 13-digit citizen registration number, which contains Korean citizens' personal information, was used as an identity verification means on line. However, this number, which was stored on firms' online databases, had been the subject of numerous thefts.[7] The Korean authorities thus decided to replace it by a new i-PIN number which does not contain any personal data and can be replaced in the event someone's i-PIN number has been copied or misused.

Relation to fraud and other crimes[8]

In most cases, perpetrators involved in this illicit activity aim at engaging in a variety of other wrongs for different purposes including obtaining credit, money, goods, services, employment benefits, or anything else of value to be used in the name of the victim, without consent.

ID thieves may sometimes not use the victims' identity themselves to commit fraud. Instead, they will sell it to other parties who will themselves commit fraud, or generate new illegal forms of personal identity (such as a birth certificate, driver's license, or a passport).

According to a 2004 US victimisation survey conducted by the Identity Theft Resource Centre (ITRC), the most prevalent form of ID theft was mostly financial (66 %); financial and criminal (9%); and financial, criminal and cloning (6 %) (ITRC, 2005, p. 6). Economic and financial fraud committed by the use of credit cards has clearly profited from technological advances. Owing to the growing number of people using electronic payment systems, this kind of fraud could propagate further in the coming years.

The online environment

Over the past 10 years, business-to-consumer (B2C) electronic transactions have increased. Numerous factors can help explain the increase: most firms in the OECD area have now a presence on the Net; in particular, financial institutions offer online banking services to their customers; finally, as consumers get more experienced in buying goods or services on line, they become more aware of how to avoid e-scams. As a consequence, hardly any business wanting to benefit from commercial opportunities can today afford to remain present solely in the offline world.

Despite its potential, the growth of the e-commerce market is somewhat restrained (OECD, 2006c).[9] This is particularly true in the European Union (EU). As noted in the *EU 2006 Special Eurobarometer* (EC, 2006, p. 12), only 27% of the EU population purchased goods and services on line over the past year, and mostly on a domestic basis. One of the main reasons explaining this limited popularity is a continuing lack of full consumer confidence in the E-marketplace. The new commercial opportunities offered by the Internet have proved to be prone to inherent risks. Those risks have taken the form of sophisticated frauds which may affect consumers and users very quickly and on a global scale, while allowing perpetrators to escape detection.

Other statistics also indicate a decline in consumer confidence in e-commerce. In a 2006 online survey conducted by the Business Software Alliance and Harris Interactive, nearly one in three adults said that security fears compelled them to shop on line less, or not at all, during the 2005-2006 holiday season (US IDTTF, 2007, Vol. I, p. 11). Similarly, a Cyber Security Industry Alliance survey in June 2005 found that 48% of consumers avoided making purchases on the Internet because they feared that their financial information might be stolen (US IDTTF, 2007, Vol. I, p. 11-12). Although the studies have not correlated these consumer attitudes with online habits, these surveys indicate that security concerns likely inhibit E-commerce.

ID theft is regarded by many as one of the major risks consumers and users are exposed to in today's digital environment. E-payment and e-banking services, on which this book will mainly focus, substantially suffer from such mistrust. In the United Kingdom, for example, an estimate of 3.4 million people were prepared to use the Internet but not willing to shop on line because of a lack of trust or fears about personal security (UK OFT, 2007, p. 6).

Many member countries have noted the problem and taken steps to help ensure that consumers and users are adequately protected against ID theft. These steps encompass various actions and measures, such as consumer and user awareness campaigns, new or adapted legislative frameworks, private-public partnerships, and industry-led initiatives aimed at putting in place technical prevention measures and responses to the threat.

Notwithstanding the initiatives outlined above, many member countries have not sufficiently addressed the problem of ID theft. As a result, its scope, magnitude and impact may vary from one country to another, or even from one source to another within the same country. These differences reflect the need for a more co-ordinated response at both domestic and international levels.

Notes

1. "A Borderless World: Realising the Potential of Global Electronic Commerce," OECD Ministerial Conference on Electronic Commerce, 7-9 October 1998, Ottawa, Canada, at *www.ottawaoecdconference.org/english/homepage.html.*
2. See Chapter 4 for a description of the terminology used in various OECD member countries as well as in international and regional organisations.
3. The concept of "personal information" refers to identification or authentication data.
4. These attributes are contained in official documents such as passports, identity card, birth and death certificates, social security numbers or driving licences.
5. This last element is a crucial means to identify individuals in the United States. This was also the case in the United Kingdom until recently. However, in early 2007, a new system of ID cards was introduced in the country.
6. Reference to this new Korean identity verification means may be found at: *www.vnunet.com/articles/print/2165834.*
7. In early 2006, 1.2 million Korean citizens noticed that their citizen registration number was used without their knowledge or consent to sign up for accounts in *Lineage*, a series of online games.
8. Discussion on the various forms of ID theft is further developed in Chapter 2.
9. See also the OECD Policy Brief on *Protecting Consumers from Cyberfraud*, OECD, Paris, October 2006, at: www.oecd.org/dataoecd/4/9/37577658.pdf, (OECD, 2006d).

Chapter 2. Online Identity Theft: Tools of the Trade

ID theft is an illicit activity with a long history that predates the Internet. Typically, conventional ID theft was – and still is – committed through techniques such as dumpster diving (also known as "bin raiding");[1] payment cards' theft; pretexting,[2] shoulder surfing;[3] skimming;[4] or computer theft.

In recent years, these activities have been given a modern spin through the fast development of the Internet, which, as examined below, allows ID thieves to install malicous software (malware) on computers, and to use a luring method known as "phishing," which itself can be perpetrated through the use of malware and spam.

ID theft based solely on malware[5]

Malware, is a general term for a software code or programme inserted into an information system in order to cause harm to that system or to other systems, or to subvert them for use other than that intended by their own users. With the rise of stealthy malware programs, such as "keystroke loggers," viruses or "Trojans"[6] that hide on a computer system and capture information covertly, malware is increasingly used as a standalone technical tool to steal victims' personal information.

ID thieves use various malware threats, including blended and targeted attacks, to obtain consumers' personal information.

Blended attacks (hidden)

Most malicious activity now combines several malware applications to carry out attacks. This has resulted in a change in landscape as blended malware attacks use techniques such as social engineering to circumvent established defences. One type of blended attack occurs for instance when malicious actors embed malware into a website that is otherwise legitimate.

Targeted attacks (hidden and/or self announcing)

Most targeted attacks notably seek to steal an entity's intellectual property and proprietary data. Because users throughout the world have taken more proactive steps to protect their systems, attackers have moved away from large scale attacks that seek to exploit as many occurrences of vulnerability as possible, to smaller more focused attacks. Targeted attacks often allow attackers to remain undetected by security measures (anti-virus software and firewalls), and maintain privileged access to a user's system for longer periods of time.

Key drivers of online ID theft: Phishing and its variants

The concept of "phishing" is not clearly and consistently defined in OECD member countries. Some law enforcement authorities or industry often use it as a synonym of ID theft. Others distinguish the two notions. As stated in the *OECD Anti-Spam Toolkit of Recommended Policies and Measures* (OECD, 2006c, p. 25), phishing is considered as the main method enabling cyber crooks to commit ID theft.

Background

Phishing is a term that was coined in 1996 by US hackers who were stealing America Online ("AOL") accounts by scamming passwords from AOL users. The use of "ph" in the terminology traces back in the 1970s to early hackers who were involved in "phreaking," or the hacking of telephone systems.

Phishing is today generally described as a luring method that thieves use to fish for unsuspecting Internet users' personal identifying information through e-mails and mirror-websites, which look like those coming from legitimate businesses, including financial institutions, or government agencies. A well-known phishing e-mail is the one pretending to be from a bank to customers asking for their account details. In France, for instance, in 2005, a one-stop-shop phishing scam targeted four banks' customers.

Another well-known phishing e-mail is the so-called "419 scam"[7] (also known as the "Nigerian scam" or offline as the "Nigerian letter"), through which phishers try to commit advance fee fraud by requesting upfront payment or money transfer from their targets. Phishers usually offer their potential victims to share with them a large amount of money that they want to transfer out of their country. Victims are then asked to pay fees, charges or taxes to help release or transfer the money. Victim of its own success, this scam is today very well known among Internet users and is not used as much anymore.

Other examples of phishing e-mails or fake websites are collected and stored by the Anti-Phishing Working Group ("APWG"), an industry association aimed at eliminating ID theft and fraud resulting from phishing. The consortium, which serves as a forum where industry, business and law enforcement agencies discuss the impact of phishing, maintains a public website allowing its members to share information and best practices for eliminating the problem.[8]

Scope

A social-engineering scheme

Phishing originally entailed deceptive attacks using deceptive or "spoofed"[9] e-mails and fraudulent websites hijacking brand names of banks, e-retailers and credit card companies, with a view to deceiving Internet users into revealing personal information (OECD, 2005, p. 23). Classic phishing attacks through e-mail can be typically described as follows:

Step 1. The phisher sends its potential victim an e-mail that appears to be from this person's bank, or some other organisation that would hold personal information. The phisher in his scam carefully uses the colours, graphics, logos and wording of an existing company.

Step 2. The potential victim reads the e-mail and takes the bait by providing the phisher with personal information by either responding to the e-mail or clicking on a link and providing the information via a form on a website that appears to be from the bank or organisation in question.

Step 3. Through this fake website or e-mail, the victim's personal information is directly transmitted to the scammer.

A technical subterfuge scheme

Like the Nigerian scam, the classic phishing attack described above is nowadays well known and fraudsters have developed more sophisticated phishing variants which are more difficult to detect, if at all possible. These variants rely on specific techniques such as malware or/and spam.

In light of the above preliminary findings, phishing may be generally described as follows:

Box 2.1 Phishing process

Phishing is perpetrated through the sending of spoofed e-mail or fake websites that trick users into disclosing their personal information and enable fraudsters to commit ID theft.

- Frequently, these messages, or the websites that they link to, try to install malicious code (OECD, 2006c, p. 25).
- Examples of the type of information sought by "phishers" include credit card and PIN numbers, account names, passwords, or other personal identification numbers.
- Phishers use victims' personal information to conduct unlawful activity, typically fraud.

Phishing techniques

Increasingly, ID theft is perpetrated using malware or "crimeware" (Radix Lab, 2006, p. 4). It is also propagated through spam messages, which often themselves contain malware.

Forms of automated phishing

Although malware is not the only means by which computers can be compromised, it provides the attacker convenience, ease of use, and automation to conduct attacks on a large scale that would not otherwise be possible due to lack of skill and/or capability. Some forms of automated phishing can be illustrated by the following attacks:

"Pharming"

Malware-based phishing attacks can take various forms. A typical malware-based phishing attack is often illustrated by the technique known as "pharming" (or "warkitting"[10]), which uses the same kind of spoofed identifiers as in a classic phishing attack, but which, in addition, redirects users from an authentic website (from a bank for instance) to a fraudulent site that replicates the original in appearance. When a user connects his/her computer to, for instance, a bank web server, a hostname lookup is performed to translate the bank's domain name (such as "bank.com") into an IP address containing several digits (such as 138.25.456.562). It is during that process that crooks will interfere and change the IP address.

"Man-in-the-middle attack"

Another example of malware-based phishing can be illustrated by the so-called "man-in-the-middle attack." This expression describes the process by which the phisher collects personal data through the interception of an Internet user's message that was intended to be sent to a legitimate site.

In today's convergent Internet, two other techniques, relying on non-computer devices, have been recently used by phishers to perpetrate ID theft.

"SMiShing"

Phishing continues to spread by reaching external devices such as mobile phones. Through this emerging technique, cell phone users receive text messages ("SMS") where a company confirms to them that they have signed up for one of its dating services and that they will be charged a certain amount per day unless they cancel their order at the company's website. Such a website is in fact compromised and used to steal the unsuspected user's personal information.

As reported by McAfee Avert Labs in August 2006, SMiShing for the first time targeted two major mobile phone operators in Spain in 2006. The scam used the two companies' own system to send texts to customers offering free anti-virus software purporting to come from the phone operator. When customers followed the link to install the software onto their computers, these were infected with malicious programs.[11] McAfee predicts that although recent, the threat of SMiShing, which is now part of the "cybercrime toolkit," will increasingly be used by malware perpetrators in the coming months (McAfee, 2006, p. 20).

"Vishing"

Voice over Internet Protocol ("VoIP") is also a new technique using phones to steal individuals' personal information. Through this means, the phisher sends a classic spoofed e-mail, disguised so as to appear from legitimate businesses or institutions, which invites the recipient to call a telephone number. Victims feel usually safer in this way as they are not required to go to a website where they would transmit their personal information. When calling, the target reaches an automated attendant, prompting her to enter personal information such as account number, password or other information for pretended "security verification" purposes. In some cases, the phisher skips the e-mail altogether and cold calls consumers, fishing for financial information.

The above described techniques based on malware are evolving and transforming into new kinds of threats on a rapid basis. They can even be mixed, as noted in 2005 by the US Identity Theft Technology Council ("ITTC") report on *Online Identity Theft* (ITTC, 2005, p. 7). The ITTC reports that "the distinctions between [phishing] attack types are porous, as many [of them] are hybrid, employing multiple technologies." The report illustrates this with the example of a deceptive phishing e-mail, which could direct a user to a site that has been compromised via content injection, and which then installs malware that poisons the user's hosts file. As a result, subsequent attempts to reach legitimate websites would be rerouted to phishing sites, where confidential information is compromised using a man-in-the-middle attack. This hybrid attack includes a pharming and man-in-the-middle attack.

Phishing vehicled through spam

Spam is another vector used for sending massive phishing hits. As described in the OECD *Anti-Spam Toolkit* (OECD, 2006c, p. 7), spam began as electronic messages usually advertising commercial products or services. Over the past few years, spam has evolved from inoffensive advertising to potentially dangerous messages which can be deceptive and may result in ID theft.

While spam messages used to be mostly text-based, they increasingly contain images. The security company adds that while spammers traditionally used well-known top level domain names such as ".com," ".biz" or .info", they now attempt to avoid detection by using domain names from small island countries, such as ".im" from the UK Isle of Man; these lesser-known domain names are often not included in spam filters (McAfee, 2006, p. 15).

Spam, phishing and malware

Although the techniques are different, spam and malware are now increasingly combined as key underpinnings of the illicit techniques fraudsters use to commit ID theft. As shown in Box 2.2 below, spam often distributes malware or direct users to infected sites aimed at creating new infected computers (OECD, 2006b).

Box 2.2 The Haxdoor example

A series of identity theft Trojan attacks were directed against Internet users in Australia between March and August 2006. Attackers used spam that contained embedded links to malicious sites to conduct the attacks. When users clicked on the link provided in the spam e-mail, they were sent to a URL containing a commercial malware kit known as WebAttacker. WebAttacker then scanned the computer system to determine which exploit would be most effective in compromising the system. Once WebAttacker determined the proper exploit, the user was directed to another webpage within the same domain and a Trojan known as "Haxdoor" was downloaded. Haxdoor then disabled various security counter-measures so that when the users visited websites, Haxdoor began capturing passwords and other data. The data captured by Haxdoor were then sent to a domain registered to the attacker. The attacker used this domain to harvest the data on a periodic basis. In one case, a computer belonging to a company in Australia was compromised by Haxdoor for approximately 14 weeks and three days. During this time, the attacker was able to access the individual's personal data including credit card details, as well as personal information for other employees in the company.

Phishing evolution and trends

Today's ID theft perpetrators are viewed as professionals. They seem to increasingly belong to organised groups which use ID theft to commit other serious crimes such as drug trafficking, money laundering, vehicle theft and illegal immigration (McAfee, 2007, p. 10; UN IEG, 2007). Their illicit actions can remain unsuspected in many cases as they are more and more relying on the help of "innocent" intermediaries, such as students.[12] Moreover, they become ever more sophisticated.

More sophisticated and tailored attacks

At the origin of the phenomenon, ID theft messages were poorly designed. They included signs of weaknesses such as: rudimentary textual errors; English-language in messages addressed to non-English individuals.

In contrast, today's phishing scams contain well-designed logos and typical messages that appear to be from real companies or institutions. To fight against this threat, in the United States, the Intellectual Property Governance Task Force[13] has been created to encourage trademark owners to employ technical measures to prevent their brands from being abused by phishers.

More targeted attacks

While continuing to target high-profile banks and e-commerce sites, phishers now try to reach fewer victims, but in more personalised ways, thereby potentially becoming even more dangerous. McAfee's 2006 *Virtual Criminality* report (McAfee, 2006, p. 11) reveals that fraudsters are changing the content of their phishing mails away from "update your details now" scams to more tailor-made messages.

They can also send false personalised e-mails to other small groups such as employees through a technique known as "spear-phishing." In such a case, the sender impersonates a colleague or the employer to steal employees' passwords and username and to ultimately gain access to the employees' company's entire computer system.

Phishing trends

Phishing is considered as a serious threat that is on the rise. A large and diverse population of fraudsters ranging from expert hackers to inexperienced individuals who can even purchase phishing kits on line have exploited the Internet and other technological tools to wreak havoc on innocent victims.[14]

In 2006, the Netcraft toolbar, an anti-phishing tool developed by the Netcraft toolbar Community,[15] blocked more than 609 000 confirmed phishing URLs, a substantive jump from only 41 000 in 2005.[16] Netcraft views this dramatic surge, mainly concentrated in November – December 2006, as the result of recent techniques implemented by phishers to automate and propagate networks of spoof pages, enabling the rapid deployment of entire networks of phishing sites on cracked web servers.[17]

The Anti-Phishing Working Group's ("APWG") November 2006 *Phishing Activity Trends* report confirms a jump in cyber attacks from July to November 2006 (APWG, 2006a). In November 2006, 37 439 new phishing sites were detected, a 90% increase since September 2006. However, in its December 2006 report (APWG, 2006b), the APWG notes a decrease in the number of new phishing sites (which dropped to 28 531).

According to the APWG's April 2007 statistics (APWG, 2007), as shown in the pie chart below, phishing websites that host malicious code or have Trojan downloads are hosted in several different countries, with the most websites hosted in the United States (38.57%), followed closely by China (37.64%) and Russia (8.87%).

What online ID thieves do with the data: credit card fraud and other abuses

Once ID thieves obtain victims' personal information, they misuse it in a variety of ways. The US Federal Trade Commission (FTC) *2006 Identity Theft Survey Report* classifies ID theft victims in one of the three main following categories:

1. New Accounts (such as new credit card, bank accounts, or loans) and Other Frauds (such as obtaining medical care);
2. Misuse of Existing Non-Credit Card Accounts; or
3. Misuse of Existing Credit Cards Only (US FTC, 2007b, p. 12).18 The most commonly reported forms of misuse of both existing credit card and non-credit card accounts perpetrated by means of ID theft were as follows (US FTC, 2007b, p. 17):

   - Credit card fraud (61%)
   - Checking or savings accounts (33%)
   - Telephone service accounts (11%)
   - Internet payment accounts (5%)
   - E-mail and other Internet accounts (4%)
   - Medical insurance (3%)
   - Others (1%)

As reflected above, credit card fraud is the most common form of misuse of existing accounts. This form of ID theft occurs when the ID thief obtains the actual credit card, the numbers associated with the account, or the information derived from the magnetic strip on the back of the card. Since credit cards can be used remotely, for example via the Internet, ID thieves are often able to commit fraud without physically possessing the victims' credit card.

ID thieves also commit new account fraud by using the victims' personal information to open an account, incur excessive charges, and then disappear. Victims often do not discover the ID theft until they are contacted by a debt collector or denied a job, loan, car, or benefit because of a negative credit rating. In some instances, ID thieves deposit stolen or counterfeit cheques, or cheques drawn on insufficient funds, and withdraw cash, causing immediate financial harm that is typically in large amounts (US IDTTF, 2007, Vol. I, p. 19-20). Although this form of ID theft is less prevalent, it can cause more financial injury, is less likely to be discovered quickly, and requires the most time for victim recovery.

Indeed, the US FTC's *2006 ID Theft Survey Report* indicates that 24% of New Accounts and Other Fraud victims did not find out about the misuse of their information until at least six months after it started – compared to just 3% of Existing Credit Cards Only and Existing non-Credit Card Account victims. In the Existing Credit Cards Only and Existing Non-Credit Card Accounts categories, the median amount of time that elapsed before victims discovered that their personal information was being misused was between one week and one month. For the New Accounts & Other Frauds category, the median value was between one and two months (US FTC, 2007b, p. 24).

Some 17% of ID theft victims reported that the thief used the victim's personal information to open at least one new account. The most common type of accounts victims reported thieves opening were as follows (US FTC, 2007b, p.19):

- Telephone service accounts (8%)
- Credit card accounts (7%)
- Loans (3%)
- Cheque / savings (2%)
- Internet payment accounts (2%)
- Auto insurance (1%)
- Medical insurance (0.4%)
- Other accounts (2%)

Some 12% of victims reported that their personal information was misused in non-financial ways (US FTC, 2007b, p.21). The most common such use, which was reported by 5% of victims, was the use of the victim's name and identifying information when the ID thief was stopped by law enforcement authorities or charged with a crime.

Other increasingly popular forms of ID theft include the misuse of personal information to obtain passports or government ID numbers (social security numbers in the United States), particularly by illegal immigrants, fraudulent requests for taxpayer refunds, and health care fraud. In particular, health care fraud may put victims at risk of receiving improper medical care due to inaccurate entries in their medical records, or having their medical insurance coverage depleted by the ID thief (US IDTTF, 2007, Vol. I, p. 20).

ID thieves may also use victims' personal information to engage in data brokering. For example, certain websites, known as "carding sites," traffic stolen credit card data all over the world. The US Secret Service estimates that the two largest current carding sites collectively have nearly 20 000 member accounts (US IDTTF, 2007, Vol. I, p. 20).

Notes

1. Dumpster diving generally refers to the act whereby fraudsters go through bins to collect "trash" or discarded items. It is the means that identity thieves employ to obtain copies of individuals' cheques, credit card or bank statements, or other records that hold their personal information.

2. Pretexting is another form of social engineering used to obtain sensitive information. In many instances, pretexters contact a financial institution or telephone company, impersonating a legitimate customer, and request that customer's account information. In other cases, the pretext is accomplished by an insider at the financial institution, or by fraudulently opening an online account in the customer's name. (See US IDTTF, 2007, Vol. I, p. 17). The US FTC has brought three cases against financial pretexters and five cases against telephone pretexters. Information on these cases is available at: *www.ftc.gov/opa/2002/03/pretextingsettlements.htm* and *www.ftc.gov/opa/2006/05/phonerecords.shtm*.

3. Shoulder surfing in relation to ID theft refers to the act of looking over someone's shoulder or from a nearby location as the victim enters her Personal Identification Number ("PIN") at an ATM machine.

4. Skimming is the recording of personal data from the magnetic stripes on the backs of a credit cards; data is then transmitted to another location where it is re-encoded onto fraudulently made credit cards.

5. Malware-related information may also be found in OECD, 2008.

6. Technical terms are defined in the Glossary.

7. This scam, which was originally committed off line, has been called "Nigerian" as, since the beginning of the 1990s, it has come primarily from Nigeria. The "419" part of the name comes from the Nigerian Criminal Code section which outlaws the practice. According to Wikipedia, the free Internet encyclopedia, as the Nigerian letter has become well known to potential targets, gangs operating it have developed variants. The targets are now often told that they are the beneficiaries of an inheritance or are invited to impersonate the beneficiary of an unclaimed estate.

8. More details about the APWG may be found in Chapter 3.

9. The term "spoof" covers any falsification of an Internet identifier such as an e-mail address, domain name or an IP address (OECD, 2005, p. 23).

10. See reference to this term at: *www.technologynewsdaily.com/node/5151.*

11. Reference to this case may be found at: *www.avertlabs.com/research/blog/?p=75.*

12. Jeevan Vasagar, *Internet criminals signing up students as "sleepers," The Guardian*, 8 December 2006, at: *www.guardian.co.uk/crime/article/0,,1967227,00.html.*

13. *See: www.ipgovernance.com/About_Us.html.*

14. See press release at: *http://fr.news.yahoo.com/12012007/7/un-kit-de-phishing-pour-les-novices-circule-sur-le.htmlb.*

15. The Netcraft toolbar Community is a digital neighbourhood watch scheme in which expert members act to defend all Internet users against phishing frauds. Once the first recipients of a phishing e-mail have reported the target URL, it is blocked for toolbar users who subsequently access that same URL. See more details at: *http://toolbar.netcraft.com/.*

16. Netcraft Toolbar Community, 2007, at: *http://news.netcraft.com/archives/2007/01/15/phishing_by_the_numbers_609000_blocked_sites_in_2006.html.*

17. These packages, known broadly as Rockphish or R11, each included dozens of sites aimed at spoofing major banks.

18. In 2003, the US FTC sponsored a first similar telephone survey of US adults, as described in its *2003 Identity Theft Survey Report* (US FTC, 2003). The aim of the survey was to estimate the incidence of ID theft victimisation, measure the impacts of ID theft on the victims, identify actions taken by victims, and explore measures that may help them in future ID theft cases. Based on the knowledge gained from the 2003 survey, FTC staff changed certain elements of the survey methodology for the 2006 survey, to more accurately capture consumers' identity theft experiences. Because of these methodological changes, estimates of the losses from ID theft in the two surveys cannot be directly compared (US FTC, 2007b, p. 8). The 2006 survey may not capture all forms of ID theft and, in particular, situations where a fictitious identity was created by combining personal information from one or more consumers with invented information may not have been captured as this form of ID theft (called "synthetic ID theft") is not always detectable by the consumer(s) whose information was used.

Chapter 3. The Impact of Online Identity Theft

Determining the impact of online ID theft is a challenging exercise. As mentioned earlier, the absence of a common legal approach to criminalising ID theft complicates matters. As a result, data measuring the extent to which ID theft can be harmful present significant weaknesses which distort the way in which the problem may be perceived. They concern a limited number of OECD member countries (including Australia, Canada, Korea, the United States, and the United Kingdom,) and may not be indicative of the situation in other countries or regions, such as the EU.

Box 3.1 Limited data on ID theft's impact on victims

- Statistics do not provide a clear picture of the notion of "victims" which either covers individuals, governments, international organisations, business and/or industry, or the economy as a whole.
- Statistics do not measure the same types of frauds or crimes and are thus incomparable.
- Statistics gathered by public authorities for policy purposes vary from those collected by private businesses for commercial purposes.
- Direct and indirect losses data do not cover all victims and all types of ID theft cases.
- Statistics do not cover all OECD member countries.

Defining the victims

Complaint data

The category of ID theft victims could *a priori* be limited to those individuals whose personal information was misappropriated by a party, and who subsequently suffered from economic or other sorts of harms. Complaints data cover this type of victim.

According to research conducted by the Canadian Fraud Prevention Forum in 2003, the impact of ID theft in Canada extended to victims across all ages, regardless of income and levels of education (BWGCBMMF, 2004, p.4). In May 2006, over 20 000 individual phishing complaints were reported in Canada, an increase of over 34% from the previous year.

The US Federal Trade Commission's (FTC) *Consumer Sentinel*[1] 2007 report on *Consumer Fraud and Identity Theft Complaint Data* (US FTC, 2007a) confirms this finding. As illustrated in Figure 3.1 below, of the people who reported their age in their ID theft complaints, young people aged between 18-29 years experienced most ID theft (29%) in 2006, followed by persons aged 30-39 (23%).

Figure 3.1 Identity theft complaints by victim age

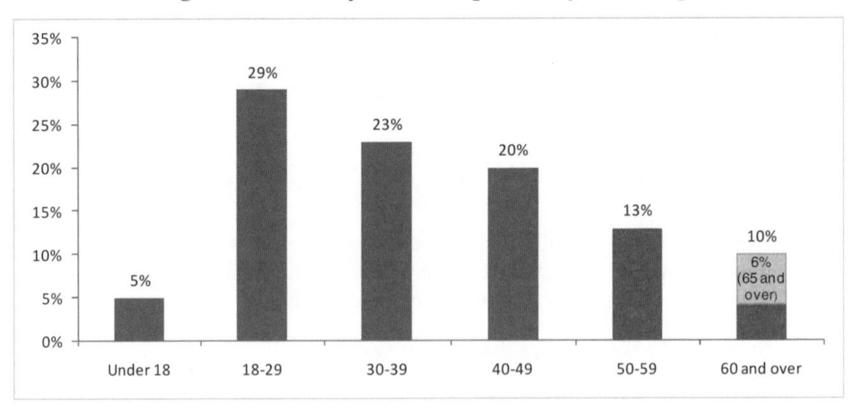

Note: Percentages are based on the total number of identity theft complains where victims reported their age (225 532). 94% of the victims who contacted the FTC directly reported their age.

Source: US FTC (2007a), Report on Consumer Fraud and Identity Theft Complaint Data.

The above complaints data only reflect reported cases, and do not illustrate the fact that in reality, the concept of victim is more complex. For example, complaint data do not always reflect that businesses and institutions are also victims of ID theft. In some instances, ID thieves may misuse a customer's banking account, or may use a bank's brand name in a phishing cyber attack to steal money from one of its customers. In such a case, the bank will also be the victim of ID theft inasmuch as it would have to reimburse the sum lost to its customer.[2]

ID theft data covering certain types of fraud

Data based on the various legal formulations of ID theft are not indicative of all ID theft cases. Since ID theft sometimes refers to a crime or is absorbed in other types of frauds, data reflecting these different realities are incomplete and do not correspond from one country to another. Consequently, data are not easy to compare, if at all possible.

According to the FTC, in 2006, for the sixth year in a row, ID theft topped the list of consumer complaints, accounting for 246 035 of more than 674 354 fraud complaints filed with the agency (see Figure 3.2 below).

Figure 3.2 Sentinel complaints by calendar year

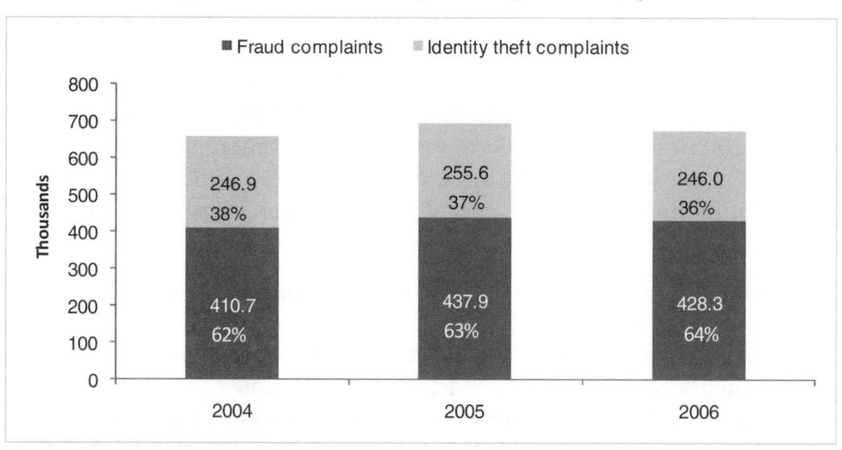

Note: Percentages are based on the total number of Sentinel complaints by calendar year. These figures exclude National Do Not Call Registry complaints.

Source: US FTC (2007a), Report on Consumer Fraud and Identity Theft Complaint Data.

The above *Consumer Sentinel* figure shows that complaints data can be classified in two categories: ID theft complaints and fraud complaints. Fraud complaints can be further broken down into Internet-related fraud complaints and other fraud complaints. However, the ID theft complaints, which are complaints involving the misuse of personal information to commit fraud, are not segregated based on whether the ID theft was perpetrated through the use of the Internet.

Divergent public and private data

There are substantial differences between statistical information gathered by public authorities for policy purposes and by private businesses for commercial purposes.

On the one hand, some security vendors state that the scale of the threat has gone down in the past years and that consumer confidence is now on the rise. Likewise, some financial institutions, which say that the costs are relatively modest, are not willing to reveal their own financial losses. On the other hand, other private bodies advance figures reflecting an increase in ID theft.

The US FTC's *2003 ID Theft Survey Report* (US FTC, 2003, p. 4) indicated that ID theft affects approximately 10 million Americans each year and that a total of 4.6% of survey participants experienced ID theft in 2002. The *2006 ID Theft Survey Report* (US FTC, 2007b, p. 4) found that approximately 8.3 million US adults discovered that they were victims of ID theft in 2005 (a total of 3.7% of survey participants).[3] In 2007, Javelin *Identity Fraud Survey Report* funded by Visa, Wells Fargo and CheckFree,[4] found that ID fraud had fallen about 12% over a year, translating into a total fraud reduction of USD 6.4 billion.

However, the Javelin report was criticised and regarded as trying to persuade the opinion that "business are doing an adequate job in protecting consumers' personal information and that the onus in on consumers to better protect themselves."[5] A 2007 McAfee survey noted this discrepancy, considering Javelin's percentages as "surprisingly low" (McAfee, 2007, p. 11) and comparing them to Gartner statistics, which, in contrast, in 2007, counted 15 million Americans as victims of ID theft.[6]

Adding to the confusion, some financial institutions even claim that none of their customers has ever been affected by a phishing attack, fearing damages to their reputation.[7]

Victims' direct and indirect losses

As mentioned above, ID theft is not a crime in itself under a vast majority of OECD member country legislation, the costs borne by victims are buried in statistics for various forms of frauds or crimes (UN IEG, 2007, p. 62). Moreover, data available alternatively refer to the costs borne by one country's economy, consumers, or consumers and business.

Victims' direct loss

In the United Kingdom, the Home Office estimates that ID fraud costs GBP 1.7 billion *to the UK economy*, compared to GBP 1.2 billion in 2002.[8] According to APACS, the UK payments association, online banking fraud continues to increase, costing the UK industry GBP 22.5 million in the first half of 2006, against GBP 14.5 million in the same period in 2005.

In the United States, according to Gartner 2007 statistics, the *average loss of funds in a case of ID theft* was USD 3 257 in 2006, up from USD 1 408 in 2005. In addition, the average loss in the opening of a fraudulent new account more than doubled over that time, from USD 2 678 to USD 5 962.[9] Based on the Javelin 2007 survey, ID theft is costing *the US industry and consumers* USD 49.3 billion annually.[10] According to a 2006 UK industry and law enforcement security report, in the United States, in 2005, 2.4 million consumers have reported losing money following a phishing attack (BT, 2006, p. 9).

In Australia, as reported by McAfee in 2007 (McAfee, 2007, p. 10), "[*estimates of the costs to the country's economy of ID fraud*] vary from less than USD 1 billion (from the Securities Industry Research Centre of Asia-Pacific) to more than USD 3 billion (Commonwealth Attorney-General's Department) per year."

According to a 2006 Computer Security Institute and FBI study surveying more than 600 US information technology ("IT") companies,[11] losses due to security incidents would have dropped by 18%, falling from USD 203 606 to USD 167 713. However, those incidents would still be significant, costing more than USD 52 billion to US IT companies.

The above variations in estimates raise the question of why data on direct losses attributed to victims of phishing vary so greatly. As observed in the *OECD Scoping Paper for the Measurement of Trust in the Online Environment*, "in part, this may be because financial institutions, while taking the threat seriously, are reluctant to publicly reveal their losses. In addition some firms may simply not know the scale of losses if they go unreported by their customers. Taken together, these factors may imply that it would be very difficult for the industry to determine a definitive figure for the direct financial losses attributable to phishing" (OECD, 2005, p. 25).

Victims' indirect loss

Victims' indirect loss can take many forms. Although there are not many statistics that clearly reflect the amount of indirect loss individual victims can suffer from ID theft, the US FTC and CIFAS have provided some guidance on this issue. According to the US FTC's *2006 Identity Theft Survey Report* (US FTC, 2007b, p. 42), 16% of ID theft victims reported having difficulty obtaining or using a credit card, 10% reported being refused a cheque account or having cheques rejected. The survey reflects that a total of 37% of all ID theft victims reported having at least one of the aforementioned problems and that 21% of victims reported having more than one of these problems.

In addition, 33% of victims who had experienced one or more of these problems said that they had spent 40 hours or more resolving problems related to their ID theft. Similarly, CIFAS reports that, victims may need between 3 and 48 hours of work to sort through problems and clear their name. CIFAS adds that "in cases where a "total hijack" has occurred, perhaps involving 20-30 different organisations, it may take the victim over 200 hours and cost up to GBP 8 000 before things are back to normal."[12]

Moreover, law enforcement authorities report that consumers victimised by ID theft may lose out on job opportunities; be denied loans for education, housing, or cars as a result of negative information on their credit reports; or be arrested for crimes they did not commit. Victims also suffer indirect financial costs, including costs incurred in lawsuits initiated by creditors and in overcoming other obstacles they face in obtaining or retaining credit. In addition, they suffer the emotional toll of having their privacy violated and the frustration of attempting to restore their reputation and credit history.

The issue of liability

ID theft can occur as a result of personal, business, or government negligence. It can also take place in instances where there is no apparent negligence from any party. How ID theft occurs has implications for liability. Businesses, for example, often due to inadequate security practices, may be exposed to the theft of sensitive consumer data they hold. As a consequence, they may bear both direct and indirect losses in the form of liability suits, fines, and loss of clientele.[13]

Both legislation and voluntary business practices deal with the issue of liability. As provided in the *1999 E-commerce Guidelines* (OECD, 1999, Section V), "[L]imitations of liability for unauthorised or fraudulent use of payments systems, and chargeback mechanisms offer powerful tools to enhance consumer confidence and their development and use should be encouraged in the context of electronic commerce." Likewise, in its 2007 *Resolution on ID theft, phishing and consumer confidence,* the Transatlantic Consumer Dialogue ("TACD") recommends "the assignment of the liability for financial damages caused by ID theft or phishing to the respective companies or service providers involved and not to consumers, unless they are proven to have acted negligently." [14] In most countries, business liability may be based on a duty for companies to provide reasonable security for all confidential data held and/or on a duty to disclose security breaches involving sensitive personal information. Such a duty, whether of a tort or contractual nature, is an important remedy tool for consumers falling victims of ID theft.[15]

In many member countries, legislation provides for ceilings of liability for the cost of fraudulent transactions by identity thieves.[16] In the United States, for example, consumer liability is limited to a maximum of USD 50 for unauthorised credit card charges; [17] for unauthorised electronic fund transfers, such liability is also limited, depending upon the timing of consumer notice to the applicable financial institution.[18] In the European Union, various instruments aim at protecting consumers from unauthorised payments. Under the 1997 European Commission's Recommendation on *transactions by electronic payment instruments and in particular the relationship between issuer and holder*,[19] consumers should bear the loss of unauthorised transactions up to a maximum of EUR 150, provided that they have not acted with extreme negligence. Under Directive 2002/65/EC *concerning the distance marketing of consumer financial services*,[20] consumers should be allowed to request cancellation of a payment where fraudulent use has been made of their payment card in connection with distance contracts, and in the event of such fraudulent use, to be re-credited with the sum paid or have it returned.[21] In the United Kingdom,[22] under the Banking Code, "[I]f someone else uses [someone's] card details without … permission for a transaction where the cardholder does not need to be present, [the cardholder] will not have to pay anything." In some countries, in addition to laws and regulations, industry practices provide consumers with "zero liability."[23]

Are there more victims off line than on line?[24]

In many instances, data does not clearly separate online ID theft cases from those occurring off line, making it difficult to assess which of the two activities is more prevalent. Whether ID theft is more dangerous on line than off line is actually a controversial issue.

Is ID theft more prevalent off line?

According to a 2006 survey by Javelin Strategy and Research ID theft is more prevalent off line than on line (JSR, 2006). Only 10% of US identity fraud cases occurred when the victim was on line, against 63% when the victim used more traditional channels. The report states that more than 90%

of identity fraud starts off conventionally by means of stolen bank statements, misplaced passwords, or bills stolen from the victim's mailbox. It concludes that if it is true that more attempts of ID theft are made on line, in only one out of 10 of those incidents did the actual successful theft of the personal data take place on the Internet. It goes on saying that this relatively low level of successful attacks is explained by the fact that a majority of leading US financial institutions have introduced adequate measures to detect and protect their customers against ID theft on line.

The conclusions of this book are however debatable. They are going against the opinion that ID theft is today more serious on line than off line.

Indeed, others have recognised the severity of the problem and noted that ID theft has a direct impact on e-commerce transactions.

According to a 2006 International Telecommunication Union online survey (ITU, 2006), more than 40% of users would refrain from transacting on line for fear that their personal information could be stolen.

Remediation tools for victims

In the United States, an important part of the multi-faceted approach to combating ID theft is developing adequate remediation tools for victims. Several federal and state laws offer victims of ID theft options to avoid or mitigate the damages caused by ID theft. Consumers are encouraged to report incidences of ID theft to the US FTC, which generates a report that may be used by law enforcement agencies to facilitate the extension of a fraud alert or to block fraudulent trade lines on a credit report.[25]

US victims of ID theft can also request one of the three Credit Reporting Agencies (Experian, Equifax, or TransUnion) to place an "initial fraud alert" on their credit report for a period of not less than 90 days, which requires creditors to confirm the consumer's identity before opening new accounts or making changes to existing accounts.[26] If a victim has an ID theft report documenting actual misuse of the consumer information, he is entitled to place a seven-year alert on his file.[27] In addition, victims may request a free copy of their credit report.[28] If the credit report contains fraudulent information as a result of the theft, the consumer may ask that the information be blocked from the credit report.[29]

ID theft victims in the United States are also advised to obtain records and application information from financial institutions that handled transactions conducted by the ID thieves in the victim's name (US IDTTF, 2007, Vol. II, p. 75).[30] In addition, US law enforcement authorities advise victims to close fraudulently opened or compromised accounts and send a letter to debt collectors claiming a debt that the victim has not incurred (US IDTTF, 2007, Vol. II, p. 76). In some states, victims may request a "credit freeze" preventing their credit reports from being released without their express consent (US IDTTF, 2007, Vol. I, p. 46).

In Australia, people can ask credit reporting agencies to place an alert on their file if they suspect they are the victim of identity theft. The Australian Standing Committee of Attorney's General 2007 *Discussion Paper on Model Identity Crime Offences*[31] also canvases the issue of assistance that may be needed for victims of ID theft.

Notes

1. The *Consumer Sentinel* database, maintained by the US FTC, contains consumer fraud complaints that have been filed with federal, state, local, and foreign law enforcement agencies and private organisations. The *Consumer Sentinel* database also houses the Identity Theft Data Clearinghouse, the US's sole government national repository of consumer complaints about ID theft. The Clearinghouse provides investigative material for law enforcement agencies and reports that provide insight into how to combat ID theft. Access to this information via *Consumer Sentinel* enables domestic and international law enforcement partners to co-ordinate their law enforcement efforts more efficiently, although not all law enforcement agencies have access to ID theft data. The US FTC's 2007 statistics are based on information from over 115 data contributors to *Consumer Sentinel*. The statistics are based on self-reporting and, therefore, understate the number of occurrences of ID theft.
2. More discussion on liabilities may be found in the later sections of this chapter.
3. The difference between the 3.7% overall prevalence figure found in the 2006 survey and the 4.6% rate in the 2003 survey is not statistically significant using standard statistical analysis (US FTC, 2007c, p.4, footnote 3). In particular, given the sample sizes and the variances within the samples, one cannot conclude on a real decrease in ID theft.
4. See reference to this survey at: *www.networkworld.com/community/?q=node/11009* or at: *www.javelinstrategy.com/2007/02/01/us-identity-theft-losses-fall-study/*.
5. See Annys Shin's article, *The Checkout*, press release, 6 February 2007, at: *http://blog.washingtonpost.com*.

6. See: *http://news.com.com/Study+Identity+theft+keeps+climbing/2100-1029_3-6164765.html.*

7. See Arnaud Devillard, *Le « phishing » en France, peu de victimes mais une menace grandissante,* press release, 10 April 2006: *www.01net.com/editorial/311785/cybercriminalite/le-phishing-en-france-peu-de-victimes-mais-une-menace-grandissante/.*

8. See: *www.identity-theft.org.uk/.*

9. See: *http://news.com.com/Study+Identity+theft+keeps+climbing/2100-1029_3-6164765.html.*

10. See: *www.javelinstrategy.com/2007/02/01/us-identity-theft-losses-fall-study/.*

11. See reference to this study at: *http://solutions.journaldunet.com/0607/060726-etude-securite-csi-fbi.shtml.*

12. See CIFAS' website at: *www.cifas.org.uk/identity_fraud_is_theft_serious.asp.*

13. See this remark from the ITRC, at: *www.idtheftcenter.org/workplace.shtml.*

14. TACD's Resolution, 8th recommendation, at: *www.tacd.org/cgibin/db.cgi?page=view&config=admin/docs.cfg&id=306.*

15. See S. L. Wood and B. I. Schecter, *Identity Theft: Developments in Third Party Liability*, at: *www.jenner.com/files/tbl_s20Publications/RelatedDocumentsPDFs1252/380/Identity_Theft.pdf.*

16. For more details on this issue, see the OECD *Report on Consumer Protections for payments Cardholders*, DSTI/CP(2001)3/FINAL, (OECD, 2001).

17. 15 U.S.C. § 1643; 12 C.F.R. § 226.12(b).

18. 15 U.S.C. 1693g; 12 C.F.R § 205.6(b). For further discussion on this issue, see Annex 4.A2 to this book.

19. Commission Recommendation 97/489/EC of 30 July 1997 *concerning transactions by electronic payment instruments and in particular the relationship between issuer and holder* (text with EEA relevance), at: *http://eur-lex.europa.eu/LexUriServ/LexUriServ.do?uri=CELEX:31997H0489:EN:NOT.*

20. Directive 2002/65/EC of the European Parliament and of the Council of 23 September 2002 *concerning the distance marketing of consumer financial services*, Official Journal L 271, 09/10/2002 p. 16-24, at: *http://eur-lex.europa.eu/LexUriServ/LexUriServ.do?uri=CELEX:32002L0065:EN:HTML.* This Directive modifies Council Directive 90/619/EEC and Directives 97/7/EC and 98/27/EC.

21. Article 8 of Directive 2002/65/EC.

22. See the UK *Banking Code*, p. 21, at: www.bba.org.uk/content/1/c4/52/27/Banking_Code_05.pdf.

23. For more details, see OECD, 2001, p. 17.

24. Philippa Lawson & John Lawford, Public Interest Advocacy Centre, Ontario, Canada, *Identity Theft: The Need for Better Consumer Protection*, November 2003, *www.travel-net.com/~piacca/IDTHEFT.pdf.*

25. US Fair Credit Reporting Act, 15 U.S.C. §§ 1681-1681x.

26. 15 U.S.C. § 1681c-1 (a)(1).

27. Id. at § 1681 c-1(b)(1)(A).

28. 15 U.S.C. § 1691 (j)(d).

29. 15 U.S.C. § 1681c-2.

30. See also 15 U.S.C. § 1681(g)(e).

31. See *www.ag.gov.au/www/agd/agd.nsf/Page/Modelcriminalcode_IdentityCrimeDiscussionPaper.*

Part II. Addressing Online Identity Theft

Part II of this book focuses on public and private efforts to combat online ID theft. Chapter 4 presents government efforts; Chapter 5 private-sector efforts; and Chapter 6 international efforts. The book concludes with suggestions for preventing online ID theft via the key areas of consumer awareness and ID authentication, as well as other measures to enhance international co-operation. The final chapter summarises key findings and recommendations.

Chapter 4. The Role of Government

Many economies do not have adequate or specific legislation prosecuting ID theft. Others, in turn, qualify this illicit activity as a crime. This difference in tackling the issue may impede consistent and effective implementation of member country co-ordinated actions to stem ID theft.

How OECD countries currently define ID theft

Only a few OECD member countries have adopted legislation that specifically addresses ID theft. In most other countries, it is a constituent element of common wrongs, and as such it is covered by a multitude of rules including unlawful access to data, fraud, forgery, and intellectual property rights, *etc.* ID theft is also a facilitating factor to commit other, more serious offences. In such a case, it is often "absorbed" by the more serious offence. The transition of traditional forms of ID theft to the electronic marketplace raises the question of whether existing OECD member countries' legislative schemes are sufficient to address the problem.

The following discussion illustrates differences in some OECD member countries' approaches.

The United States: ID theft is a specific offence

In the United States, under Federal and various state laws, ID theft is a specific criminal offence that can occur when someone "knowingly transfers, possesses, uses, without lawful authority, a means of identification of another person with the intent to commit, or in connection with, any unlawful activity that constitutes a violation of Federal law, or that constitutes a felony under any applicable State or local law." [1] As such, ID theft is a crime *per se*. In 2004, the *Identity Theft Penalty Enhancement Act* ("ITPEA") introduced aggravated penalties.[2]

United Kingdom: ID theft is a constituent element of other wrongs or crimes

Until recently, fraud in the United Kingdom was not regarded as a specific offence. However, under the UK *Fraud Act 2006*, which came into force on 15 January 2007,[3] fraud has become a single statutory offence which extends to fraud committed on line. Fraud can be committed in the three following ways: *i*) by making a false representation (dishonestly, with intent to make a gain, cause loss or risk of loss to another); *ii*) by failing to disclose information; and *iii*) by abuse of position. In addition, new offences were established such as: the act of "obtaining services dishonestly … if payment is made for them," such as credit card fraud over the Internet; the act of possessing "articles for use in frauds" (the term "article" including "any program or data held in electronic form"), which relates to ID fraud; and the act "for making or supplying articles for use in frauds… knowing that it is designed or adapted for use … in connection with fraud," which relates to the writing of malicious software.[4]

Australia: Not a stand-alone offense

In Australia, except in Queensland[5] and South Australia,[6] ID theft is not a stand-alone offence. However, identity crime and identity security have increasingly become a priority for the Australian Government since 2000, with the tabling of the *Numbers on the Run* report by the House of Representatives Standing Committee on Economics, Finance and Public Administrations. In July 2004 the Australian Standing Committee of Attorneys General ("SCAG") agreed to the development of model identity theft offences. A Discussion Paper on model identity crime offences was developed and approved by SCAG at their April 2007 meeting.[7]

Canada: Proposal to make ID theft criminal offense

On 21 November 2007, the Government of Canada introduced a new bill making ID theft a specific criminal offence. While the misuse of someone's identity information is covered in the *Criminal Code* by offences such as impersonation and forgery, the preparatory steps of collecting, possessing and trafficking in identity information are generally not captured by existing offences.[8]

Box 4.1 Survey of Terminology Used to Define ID Theft

"ID Theft", "ID Fraud" or "ID Crime"?

OECD member country law enforcement authorities, industry, academics and the media use a variety of terminologies to characterise the problem. In some countries, such as in the United States, Canada, or Korea, the concept of ID theft prevails. In contrast, some other countries, mostly in the European Union, use the terms "identity fraud" or "identity crime" as synonyms. French-speaking countries refer to the terms ID theft ("vol d'identité") or impersonation ("usurpation d'identité") interchangeably.

According to Europol's 2006 EU Organised Crime Threat Assessment ("OCTA"), ID theft and ID fraud are subsets of ID crime. ID theft is itself a sub-category of ID fraud, as "identity fraud is broader than identity theft in that identity fraud refers to the fraudulent use of any identity, real or fictitious, while identity theft is limited to the theft of a real person's identity" (Europol, 2006, p. 18).

For the UN Intergovernmental Expert Group to prepare a study on Fraud and the Criminal Misuse and Falsification of Identity ("the UN IEG") (UN IEG, 2007), ID theft and fraud are also sub-categories of ID crime, which covers all forms of illicit conduct involving identity. It considers that cases in which identities or related information were simply fabricated are not analogous to either fraud or theft. The UN IEG distinguishes ID theft from ID fraud as follows:

- *Identity theft* refers to occurrences in which information related to identity, which may include basic identification information and in some cases other personal information, is actually taken in some manner analogous to theft or fraud, including theft of tangible documents and intangible information, the taking of documents or information which are abandoned or freely available, and the deception of persons who have documents or information into surrendering them voluntarily.
- *Identity fraud* refers to the use of identification or identity information to commit other crimes or avoid detection and prosecution in some way.

According to the UN IEG, the element of deception, and hence the term "fraud," lies not in the use of deception to obtain the information, but in the subsequent use of the information to deceive others. As with economic fraud, this element of deception includes the deception of technical systems as well as human beings.

In the United Kingdom, ID theft has been defined (Home Office Identity Fraud Steering Committee, 2006) as "the act whereby someone obtains sufficient information about an identity to facilitate identity fraud ("ID fraud"), irrespective of whether, in the case of an individual, the victim is alive or dead." ID theft, whether on or off line, is therefore a preparatory step to commit subsequent fraud.

The European Union: No specific provisions

To date, except for the United Kingdom, no specific provisions about ID theft or ID fraud may be found in EU Member States' laws. In France, in 2005, a legislative proposal was put forward before the Senate (high chamber of the French Parliament) to specifically prosecute online ID theft. However, in October 2006, the French Minister of Justice rejected this proposal[9] stating that ID theft (whether committed on or off-line) is adequately sanctioned under French law through various criminal and civil wrongs such as the use, without authorisation, of someone's name, if such a use exposes the victim to criminal sanctions (Article 434-23 of the French Criminal Code), fraud ("*escroquerie*" – Article 313-1 of the French Criminal Code), or defamation (Article 131-12 *et al.* of the French Criminal Code).

The option of criminalising ID theft

Would criminalisation improve OECD member countries' fight against ID theft? As described above, some countries argue that even though ID theft is not an offence *per se* under their legislation, it is already adequately covered by various provisions relating to data privacy, security or frauds, which may be of a criminal nature. Others however consider the option of criminalisation as a helpful means to better stem the threat.

The UK Credit Industry Fraud Avoidance Service ("CIFAS") sees "the criminalisation of ID theft through sections 25 and 26 of the UK *Identity Cards Act 2006* and through the Fraud Bill" as a progress. CIFAS takes the view that, before criminalisation, fraudsters could well perceive "ID fraud as an excellent avenue to financial gain precisely because they knew that the police was unlikely to pursue them" (CIFAS, 2006, p. 5). It considers that this will create incentives for all stakeholders to understand the specificity of the problem: the Police will intervene and investigate more ID theft cases; the victims will receive a voice and the needed recognition that they are victims of a crime (INTERVICT, section 7.3); penalties will be increased to address the issue more aggressively.

In Canada, the proposed Identity Theft bill aims at filling a gap in the *Criminal Code,* to help ensure that the preparatory steps of collecting, possessing and trafficking in identity information were generally not captured by existing offences. Under the new legislation, each of the three offences created is subject to five-year maximum sentences.

The US Identity Theft Task Force supports the effort to encourage other member countries to enact suitable legislation criminalising ID theft. Since

ID theft is a global problem, providing assistance to, and receiving assistance from, foreign law enforcement agencies is critical. In cases where the foreign country does not have laws criminalising ID theft, it impedes the investigating country's ability to collect evidence and prosecute ID theft crimes that have foreign components (US IDTTF, 2007, Vol. I, p. 59-60).

In addition to enacting legislation that creates specific criminal offences for ID theft, OECD member countries could adopt a multi-faceted approach including, for example, legislation on data security breach notification[10] and public-private initiatives to develop a comprehensive solution to the ID theft problem.

Public education and awareness campaigns

Business and industry sometimes claim that individuals should be more aware of their responsibility to contribute to safeguarding their identities. However, individuals will not retain effective control over their personal data without strong education efforts. Internet users' awareness about ID theft should therefore be a key priority for governments, business and industry (OECD, 2006b, p. 74). Some studies have indeed shown that users do not yet fully realise the extent to which they can expose themselves to online threats.[11]

Evidence from the 2006 EU Barometer suggests that consumer education, information and empowerment play a key role in raising consumer confidence and boosting cross-border transactions. Throughout the survey, it has been apparent that the more educated and informed consumers are, the more confident they feel in making cross-border purchases. Likewise, the 2003 Cross-border Fraud Guidelines recommend that member countries "educate consumers about fraudulent and deceptive commercial practices" (OECD, 2003, Section II. F). As regards ID theft, education should not be limited to consumers. All users (including all actors, whether public or private, who participate in the enforcement of anti-ID theft rules and practices) should be aware of the problem, how it occurs, who it affects and how it can be avoided or stopped.

Sample consumer-education campaigns

Many countries have launched education and awareness campaigns targeting consumers - and more generally users - to alert them about the risks of becoming a victim of ID theft and to help them address the problem once it has occurred (see also the Annex at the end of this chapter).

Australia: In February 2004, the Australian Government developed and released a national *Identity Theft Information Kit*, to provide the Australian community with information and strategies on how to prevent and respond to identity theft. The Australian Government also released a concise information booklet on identity theft in March 2007 as part of the Australasian Consumer Fraud Taskforces (ACFT) identity theft week, which operated as part of the ACFT's annual fraud awareness campaign.[12]

United Kingdom: The Home Office Identity Fraud Steering Committee launched the *www.identity-theft.org.uk* website which contains tips about how to avoid ID theft and where victims can find technical and legal help and support to clear out their misused identity. The website notably contains a link to the *GetSafeOnline* website which was created to support "*GetSafeOnline*," the first UK Internet security awareness campaign launched in 2005.[13]

A 2006 *GetSafeOnline* report revealed the first positive results of the campaign. According to a market research launched in October 2005, 62% of people had identified and understood the campaign messages, resulting in 30% keeping their personal details safe. However, the report points out remaining challenges showing a "knowledge security gap" since 40% of people remain unsure of where to go to get advice on how to stay safe on line (*GetSafeOnline*, 2006, p. 10). The UK Department of Trade and Industry ("DTI") has likewise developed "*Consumer Direct*," a telephone and Internet advice service providing consumers with information on fraud and notably on phishing.[14]

Mexico: The public University of Mexico ("UNAM") has put in place various websites to alert consumers and users about all sorts of security risks on line. Advice on the identification of scams (*www.seguridad.unam.mx/doc?ap=articulo&id=121*), pharming (*www.seguridad.unam.mx/usuario-casero/pharming.dsc*), phishing (*www.seguridad.unam.mx/usuario-casero/phishing.dsc.*) and tips to prevent privacy (*www.seguridad.unam.mx/doc?ap=articulo&id=118*) and security breaches are provided to users.

Canada: PhoneBusters National Call Centre ("PNCC"),[15] the country's anti-fraud call centre operated by the Royal Canadian Mounted Police and the Ontario Provincial Police, educates consumers on fraud scams through its nationwide awareness campaign "*Recognise It. Report It. Stop It.*" The campaign includes tips on how to avoid falling victim to identity theft and where to report it.

United States: In February 2005, the Federal Trade Commission organised, in co-operation with federal, state and local agencies, and national advocacy organisations, the National Consumer Protection Week ("NCPW"), during which it educated consumers about the theme of "Identity Theft: When Fact Becomes Fiction." The 2005 NCPW focused on minimising the risk of ID theft and taking fast action in the event it occurs. The FTC also launched the NCPW website, *www.consumer.gov/ncpw*, which contains information for consumers and businesses including phishing.

On 10 May 2006, the FTC launched another nationwide campaign to educate consumers about ID theft.[16] Starting from the premise that there is no guarantee that ID theft can be avoided, the "*Deter, Detect, Defend*" campaign aims at helping consumers take steps to minimise the risks and damages if the problem occurs.[17] The campaign is conducted through the distribution of brochures, kits,[18] and a warning video.[19]

The US Department of Justice ("US DOJ") is also active in educating consumers about ID theft and has produced radio and television public service announcements about ID theft (US IDTTF, 2007, Vol. II, p. 30-31). The Federal Deposit Insurance Corporation ("FDIC") has also prepared educational tools and is working on an educational campaign, scheduled for rollout in 2007, to educate consumers about online banking and the protections available to them that make it safe. The US Social Security Administration, the US Postal Inspection Service ("USPIS"), the US Department of Education, the US Department of Health and Human Services, and the US Securities and Exchange Commission ("SEC") have also published guidance to consumers on ID theft.

In some other OECD countries, however, public information on how to avoid being caught by cyber fraudsters, where to report and file a complaint is not always available.

Empowering domestic enforcement

Enhanced domestic co-ordination

In the United Kingdom, the Government has established the Identity Fraud Steering Committee, which includes representatives of a wide range of public agencies, law enforcement and the private sector. It benefits from CIFAS, a private organisation run by the financial sector which receives reports of attempted or completed frauds against the financial sector. Together with UK credit reference agencies, CIFAS assists the financial sector in verifying the identity of individuals applying for financial services, and informs industry whether that individual has been linked to previous frauds.

In the United States, various Federal authorities are empowered to investigate and prosecute ID thieves. The US DOJ and federal prosecutors work together with federal investigative agencies such as the *Federal Bureau of Investigation* ("FBI"), the *US Secret Service*, and the *United States Postal Inspection Service* to prosecute ID theft and fraud cases. These agencies have launched special enforcement initiatives to combat ID theft. For example, the FBI Cyber Division also conducted *Operation Spam Slam*, a criminal spam and malicious code investigative initiative supported daily by more than 20 small and medium-sized enterprises. The private sector also provided additional support for the project, which has resulted in more than 100 investigations (US IDTTF, 2007, Vol. II, p. 50-51).

US state criminal law enforcement efforts are also critical to the fight against ID theft. All 50 states and the District of Columbia have some form of legislation that prohibits ID theft. According to a 2005 National Survey of State Court Prosecutors (US IDTTF, 2007, Vol. II, p. 45), 69% of all prosecutors surveyed, and 97% of prosecutors surveyed from areas with populations of 1 million or more, had litigated at least one computer-related ID theft case.[20] In addition, 80% of all prosecutors surveyed, and 91% of prosecutors surveyed from areas with populations of 1 million or more, had litigated a computer-related credit-card fraud case.

A number of US federal and state law enforcement authorities have formed multi-agency task forces or working groups to facilitate the prosecution of ID theft laws. US federal authorities, including the US Attorney's Offices, FBI, Secret Service, and USPIS, lead or co-lead more than 90 task forces and working groups devoted (in whole or in part) to ID theft.

In May 2006, the US Office of Community Oriented Policing Services ("COPS") presented its national strategy to stem ID theft through several recommendations including the creation of state-level co-ordination centres providing crime analysis, victim assistance, state-wide investigations and law enhanced enforcement co-operation.

In addition, on 10 May 2006, the US Identity Theft Task Force, comprised of 17 federal agencies and co-chaired by Attorney General Alberto R. Gonzales and Federal Trade Commission Chairman Deborah Platt Majoras, was created.[21] The Task Force aims at developing a nationwide strategic plan to better prevent ID theft, co-ordinate prosecution, and ensure recovery for the victims. One of the main goals of the Task Force is to promote enhanced co-operation by Federal departments and agencies with state and local authorities responsible for the prevention, investigation, and prosecution of ID theft, to notably avoid unnecessary duplication of effort and expenditure of resources. In January 2007, the Task Force launched a public consultation (US FTC, 2007a) to receive comments on its Interim Recommendations (US FTC, 2006) on ways to improve the effectiveness and efficiency of federal government efforts to reduce ID theft.

On 23 April 2007, the Task Force issued a report that sets forth a strategic plan for addressing myriad challenges presented by ID theft. The strategic plan focuses on four key areas: "(1) keeping sensitive consumer data out of the hands of identity thieves through better data security and more accessible education; (2) making it more difficult for identity thieves who obtain consumer data to use it to steal identities; (3) assisting the victims of identity theft in recovering from the crime; and (4) deterring identity theft by more aggressive prosecution and punishment of those who commit the crime." (US IDTTF, 2007, Vol. I, p. 4). The Task Force made a number of recommendations in each of these four areas, including reducing unnecessary use of Social Security Numbers, establishing national standards for private sector entities to safeguard the personal data they collect and to provide notice to consumers when a breach occurs that poses a significant risk of ID theft, implementing sustained consumer awareness campaigns, and establishing a National Identity Theft Law Enforcement Center to enhance law enforcement co-operation in prosecuting ID thieves.

The public and the private sectors have collaborated to improve information sharing on ID theft. For example, the US Secret Service hosts a portal called the e-Information system for members of the law enforcement and banking communities (US IDTTF, 2007, Vol. II, p. 57). The e-Information system provides a forum for members to post the latest information on scams, counterfeit cheques, frauds and swindles, and updates Bank Identification Numbers. The USPIS created the Intelligence Sharing Initiative website in 2005, which allows fraud investigators representing retail and financial institutions to share information pertaining to mail theft, ID theft, financial crimes, investigations, and prevention methods. The website also gives users access to the "Hot Addresses List"– a list of addresses located throughout the United States and Canada linked to a variety of fraud schemes, including fraudulent applications schemes, account takeover schemes, mail order schemes, and reshipping schemes (US IDTTF, 2007, Vol. II, p. 58).

In Japan, the Council of Antiphishing Japan,[22] a private forum composed of 22 companies and 3 associations, collects data and trends about phishing scams and provides technical and legal advice on how to stem ID theft. Japan's Ministry for Economy, Trade and Industry ("METI"), Ministry of Internal Affairs and Communications ("MIC") and the National Police Agency ("NPA") hold a status of observer in the Council's activities. In addition, MIC hosts meetings of the "Contact Group to Promote Countermeasures Against Phishing" with related businesses and government offices to share information between ISPs and to search for effective countermeasures. METI and NPA hold the status of observer in the Contact Group's activities.

Resources and training

Many member country enforcement authorities often lag behind cyber fraudsters to keep up-to-date with technology advances and the methods to protect them. These actors should receive appropriate technical education to ensure that they are aware of how thieves operate. Such education should be done through regular training to ensure that authorities are kept up-to-date with the rapid evolution of ID theft techniques. More resources should be devoted in this respect (NCL, 2006; UN CC IEGFCMFI, 2007).

The US ID Theft Task Force has recommended that US federal agencies assist, train and support foreign law enforcement through the use of Internet intelligence-collection entities, including the Internet Crime Complaint Center and the Cyber Initiative Resource Fusion Center, as well as engage in joint investigations (US IDTTF, 2007, Vol. I, p. 62). On the national level, US Attorney's Offices participate in ongoing training seminars, and several law enforcement agencies – including the US DOJ, USPIS, the US Secret Service, the US FTC, and the FBI – along with the American Association of Motor Vehicle Administrators, have jointly sponsored over 20 regional, one-day training seminars on ID theft for state and local law enforcement agencies across the country (US IDTTF, 2007, Vol. II, p. 71-73).

Several other agencies actively co-ordinate training programmes. Notably, the US Secret Service has developed a training video on ID theft for police departments, and its Electronic Crimes Section has trained over 150 state and local officers from across the United States to conduct computer investigations and computer forensic analysis (US IDTTF, 2007, Vol. II, p. 72). The National White Collar Crime Center ("NW3C"), a non-profit organisation that provides training programmes and other assistance to state and local law enforcement authorities in partnership with the Bureau of Justice Assistance, has developed a three-day ID theft course. The curriculum includes topics such as tools, techniques, and resources for investigating ID theft crimes, and the basics of ID theft for financial gain.

Disclosure of data security breaches

In the United States, the *Gramm-Leach-Bliley Act* ("GLB Act") introduced mandatory protection of consumers' personal financial information by financial institutions.[23] The GLB Act addresses two distinct types of protection for personal information: security and privacy. The security provisions require the agencies to write standards for financial institutions regarding appropriate physical, technical, and procedural safeguards to ensure the security and confidentiality of customer records and information, and to protect against anticipated threats and unauthorised access to such information.[24]

The privacy provisions require financial institutions to notify their customers of their information-sharing practices and provide customers with an opportunity to opt out of information-sharing with certain unaffiliated third parties in certain circumstances.[25] Various federal agencies, including the federal bank regulatory agencies, the FTC, and the Securities and Exchange Commission ("SEC"), have issued regulations or guidelines addressing both the security and privacy provisions of the GLB Act.[26]

The US states have also recognised the importance of breach notification. In 2003, California passed two significant data security breach laws.[27] Likewise, 37 US states have laws that impose a duty on companies to disclose security breaches affecting sensitive personal information of their customers, and some laws have safeguard and disposal requirements (US IDTTF, 2007, Vol. I, p. 34). The states have taken a variety of approaches regarding when notice to consumers is required. Some states require notice to consumers whenever there is unauthorised access to sensitive data. Other states require notification only when the breach poses a risk to consumers, and do not require notice when the data cannot be used to commit ID theft, or when technological protections prevent access to the data. A number of bills have been introduced in the US Congress that would establish a federal notice requirement.

In 2005 and 2006, the US FTC brought 14 enforcement actions against companies for failure to provide reasonable data security (US IDTTF, 2007, p. 11). Its action against ChoicePoint, a US-based company compiling and selling consumers' personal information to more than 50 000 businesses, was largely covered in the press. In 2005, ChoicePoint recognised that it had sold information about more than 163 000 consumers to people who turned out to be ID thieves. On that basis, the US FTC charged the company for violating the US Fair Credit Reporting Act by furnishing consumer reports to subscribers who did not have permission to obtain them, and by failing to maintain reasonable procedures to verify both their identities and how they intended to use the information. The company agreed to reimburse the victims of ID theft and committed to: implement new procedures to ensure that only legitimate businesses would be provided with consumer reports; establish and maintain a comprehensive information security program; and obtain audits by an independent third-party security professional every other year until 2026.[28]

Most companies are reluctant to disclose data security breaches, fearing a negative impact on their business. The 20% decline that ChoicePoint incurred in its stock prices following its 2005 disclosure illustrates this risk.

Japan: data breach notification is considered in the context of the Japanese Act on the Protection of Personal Information, which went into effect in 2005. More specifically, the Cabinet Office of Japan has issued a "Basic Policy on the Protection of Personal Information"[29] stating that, in the case of leakage of personal information, it is important that the concerned entity make public the fact of the case to the extent possible, in order to prevent secondary damage and avoid occurrence of similar cases. Under the Guidelines issued by the Financial Services Agency, financial institutions are under the duty to report any data leak to the authorities.

Australia: the National Privacy Principles ("NPPs") and Information Privacy Principles ("IPPs") in the *Privacy Act 1988 (Cth)* impose privacy obligations on government agencies and private organisations in relation to the handling of personal information. In particular, NPP's and IPP's require that agencies and organisations must take reasonable steps to protect the personal information they hold from misuse and loss and from unauthorised access, modification or disclosure. Agencies and organisations must therefore ensure that they have appropriate security requirements in place to ensure that breaches do not occur and must be accountable for when breaches occur.

In the context of its current review of privacy laws, the Australian Law Reform Commission ("ALRC") is considering the issue of whether Australia should introduce requirements in the *Privacy Act 1988 (Cth)* to require organisations and agencies to notify individuals of data breaches. On 12 September 2007, the ALRC released *Discussion Paper 72: Review of Australian Privacy Law*, which outlines draft proposals for reform. The ALRC is due to provide a final report to Government by the end of March 2008.

EU member dtates: There are currently no data breach disclosure obligation in EU member states. Article 4 of the 2002 Directive on the Processing of Personal Data and the Protection of Privacy in the Electronic Communications Sector ("the e-Privacy Directive")[30] requires ISPs to notify their subscribers about security risks, without however imposing on them to notify actual security breaches. Under the European Commission's proposed reform of the regulatory framework for electronic communications,[31] ISPs would be required to notify both the national regulatory authorities and their customers of any breach of security that led to the loss of personal data and/or to interruptions in the continuity of service supply and their customers of any breach of security leading to the loss, modification or destruction of, or unauthorised access to, personal customer data. The proposal has been welcomed by the Article 29 Data Protection Working Party.[32] In its comments on the proposal in December 2006, the US FTC invited the European Commission to carefully consider whether it should require service providers to notify customers of security breaches under all circumstances. The US FTC rather suggested that consumers should be notified in the event of a security breach creating a significant risk of identity theft or other related harm.[33]

Notes

1. See the United States Code ("U.S.C."), Title 18, Section 1028 (a) (7).
2. The text of the ITPEA is available at: www.consumer.gov/idtheft/pdf/penalty_enhance_act.pdf.
3. The UK *Fraud Act 2006* is available at: www.opsi.gov.uk/acts/acts2006/ukpga_20060035_en.pdf.
4. Out-Law news, *Phishing kits banned by new Fraud Act*, 13 November 2006, at: *www.out-law.com/page7469.*
5. Queensland *Criminal Code and Civil Liability Amendment Act 2007.* The Act was enacted in March 2007.
6. South Australia *Criminal Law Consolidation ("Identity Theft") Amendment Act 2003.* The Act came into force on 5 September 2004.
7. More details about the SCAG model identity theft offences can be found at: *www.ag.gov.au/www/agd/agd.nsf/Page/Modelcriminalcode_IdentityCrimeDiscussionPaper.*
8. See: *http://canada.justice.gc.ca/en/news/nr/2007/doc_32178.html.*
9. Marc Rees, *L'usurpation d'identité sur Internet n'aura pas sa loi*, 23 October 2006, at: *http://fr.news.yahoo.com/23102006/308/l-usurpation-d-identite-sur-Internet-n-aura-pas-sa.html.*
10. Discussed later in this chapter.
11. In the United States, according to the *Consumer Reports Journal* of September 2006 (see also www.ConsumerReports.org), 20% of the households surveyed in 2005 and 2006 did not have anti-virus software installed on their computers. In the United Kingdom, according to an October 2006 *GetSafeOnline Report* (*GetSafeOnline*, 2006), one fifth of a survey respondents had not updated their virus protection in the last month and 23% had opened an e-mail attachment from an unknown source.

12. Copies of the Identity Theft Information Kit and Booklet are available at: *www.ag.gov.au/www/agd/agd.nsf/Page/Crimeprevention_Identitysecurity#q3*.
13. *GetSafeOnline* is a private-public initiative between the UK Government, the Serious Organised Crime Agency and companies including BT, eBay.co.uk, HSBC, Microsoft and SecureTrading.
14. See: *www.consumerdirect.gov.uk/watch_out/scams/phishing/*.
15. The PhoneBusters campaign is available at: *www.rcmp.ca/scams/brochure_e.pdf*.
16. The US FTC's education campaign may be found at: *www.ftc.gov/bcp/edu/microsites/idtheft//*.
17. See the press release announcing the campaign at: *www.ftc.gov/opa/2006/05/ddd.htm*.
18. The consumer education kit aims at helping institutions educate their employees and customers to minimise their risk. It includes a victim recovery guide, a training booklet, a guide to talk about ID theft, a brochure and a 10 minute video.
19. The warning video is available at: *www.ftc.gov/bcp/edu/microsites/idtheft/law-enforcement/index.html*.
20. Bureau of Justice Statistics Bulletin, Prosecutors in State Courts, 2005, at 5 July 2006, available at: www.ojp.usdoj.gov/bjs/pub/pdf/psc05.pdf.
21. Exec. Order No. 13, 71 FR 27945 (May 10, 2006), available at: *www.whitehouse.gov/news/releases/2006/05/20060510-3.html*. More information is available on the Task Force website, at: *www.idtheft.gov*.
22. Details on the Council's activities may be found at: *www.antiphishing.jp/index.html*.
23. 15 U.S.C. § 6801-09. In addition to the GLB Act, a variety of other US laws impose data security requirements for particular entities in certain contexts. They include Section 5 of the FTC Act, 15 U.S.C. § 45(a), which prohibits unfair or deceptive acts or practices; the Fair Credit Reporting Act, 15 U.S.C. §§ 1681-1681x, which restricts access to consumer reports and imposes safe disposal requirements, among other things; the Health Insurance Portability and Accountability Act of 1996 (HIPAA), 42 U.S.C. § 1320d et seq., which protects health information; section 326 of the USA Patriot Act, 31 U.S.C. § 5318(1), which requires verification of the identity of persons opening accounts with financial institutions; and the Driver's Privacy Protection Act of 1994, 18 U.S.C. § 2721 et seq., which prohibits most disclosures of drivers' personal information. (US IDTTF, 2007, Vol. I, at 31 & notes 42-48).
24. 15 U.S.C. §§ 6801-09.
25. Idem.

26. 16 C.F.R. Part 313 (FTC); 12 C.F.R. Part 30, App. B (OCC, national banks); 12 C.F.R. Part 208, App. D-2 and Part 225, App. F (FRB, state member banks and holding companies); 12 C.F.R. Part 364, App. B (FDIC, state non-member banks); 12 C.F.R. Part 570, App. B (OTS, savings associations); 12 C.F.R. Part 748, App. A (NCUA, credit unions); 16 C.F.R. Part 314 (FTC, financial institutions that are not regulated by the FRB, FDIC, OCC, OTS, NCUA, CFTC, or SEC); 17 C.F.R. Part 248.30 (SEC); 17 C.F.R. Part 160.30 (CFTC). The US ID Theft Task Force's report includes a discussion about these regulations (US IDTTF, 2007, Vol. II, at 1-6).

27. California was the first US state which introduced such legislation. Under the California *Security Breach Information Act* (which entered into force on 1 July 2003), any company storing customer data electronically should notify its California customers of a security breach to the company's computer system when the company knows or has reason to believe that unencrypted information about customers has been disclosed (S.B. 1386 and see California *Civil Code*, Section 1798.82). In addition, the California *Financial Information Privacy Act* sets out limits on the ability of financial institutions to share non-public personal information about their customers with affiliates and third parties.

28. See: www.ftc.gov/opa/2006/12/choicepoint.htm.

29. Following the Cabinet Office's Decision dated 2 April 2004.

30. EU Parliament and Council Directive 2002/58/EC on the Processing of Personal Data and the Protection of Privacy in the Electronic Communications Sector, 12 July 2002, in *Official Journal of the European Union* ("OJEU"), L 201/37, at: *www.spamlaws.com/docs/2002-58-ec.pdf*.

31. Communication from the Commission to the European Council, Parliament, Economic and Social Committee and Committee of the Regions on the *Review of the EU Regulatory framework for electronic communications networks and services*, SEC(2006) 816 28 June 2006, paragraph 7.2, p. 29, at: *http://ec.europa.eu/information_society/policy/ecomm/doc/info_centre/public_consult/review/staffworkingdocument_final.pdf*.

32. See the Article 29 Data Protection Working Party's *Opinion 8/2006 on the review of the regulatory framework for Electronic Communications and Services, with focus on the ePrivacy Directive*, 26 September 2006, at: *http://ec.europa.eu/justice_home/fsj/privacy/docs/wpdocs/2006/wp126_en.pdf*.

33. See: *http://useu.usmission.gov/Dossiers/Internet_Telecoms/Dec2006_FTC_Comments.asp*.

Annex 4.1 ID Theft: Education and Government Initiatives in OECD Countries

Selected governments

Australia

The Australian Government distributes an information kit, *How to prevent and respond to identity theft* (www.crimeprevention.gov.au), to provide the community with practical strategies on how to avoid becoming a victim of ID theft. In 2007, it released a brochure, *ID Theft: Dealing with identity theft*, as part of the Australasian Consumer Taskforce's Identity Theft Week, which operated as part of the Taskforce's annual fraud awareness campaign.

The government also distributes a booklet, *E-Crime - A Crime Prevention Kit for Small Business*, which is aimed at helping small business owners identify what to do to avoid becoming a victim of e-crime. In July 2007 the government introduced a range of e- security initiatives under the E-Security National Agenda. These include initiatives to raise awareness of e-security among home users and small business and the expansion of the national and international e-security exercise program. The government's e-security website, *Stay Smart Online* (www.staysmartonline.gov.au), provides online users with practical tips on how to secure a personal computer, smart transacting on line, and information on keeping children and young people safe on line. The Australasian (Australia and New Zealand) Consumer Fraud Taskforce maintains *ScamWatch* (www.scamwatch.gov.au), a consumer scam information website providing information on several types of scams, schemes and fraud. It also provides the facility for reporting scams.

Belgium

In Belgium, various education campaigns against Internet threats, including ID theft, are run on all kind of supports such as guides ("Guide for the Internet user"), websites (www.saferinternet.be, which targets children, http://economie.fgov.be – of the Federal Public Service Economy and which contains information on consumers' rights under Belgian laws), press releases on Internet fraud to draw consumers' attention to Internet fraud practices, such as phishing.

Canada

The Consumer Measures Committee ("CMC"), an organisation representing federal, provincial and territorial Ministries responsible for Consumer Affairs, has developed an information kit to help consumers avoid becoming victims of identity theft, and to provide guidance on procedures to take if they do. In addition, CMC has developed a guidance document for businesses, providing them with tips on how to protect their customers' personal information (see: www.cmcweb.ca/idtheft).

A number of other initiatives are carried out to inform consumers about ID theft. These include the Fraud Prevention Forum, a group of government, law enforcement and private sector organisations, which leads the annual Fraud Prevention Month campaign every March under the theme *Fraud: Recognize it. Report it. Stop it.* Identity theft, as a type of fraud, makes up a significant portion of the information that is presented to the public during Fraud Prevention Month.

Japan

In Japan, the Ministry of Internal Affairs and Communications (MIC) launched an *Information Security Site for General Users* (www.soumu.go.jp/joho_tsusin/ security/index.htm), a website that provides basic information on information security including countermeasures to combat online threats such as ID theft.

Mexico

In Mexico, the public University of Mexico ("UNAM") has put in place various websites to alert consumers and users about all sorts of security risks on line. Advice on the identification of scams (*www.seguridad.unam.mx/doc?ap=articulo&id=121*), pharming (*www.seguridad.unam.mx/usuario-casero/pharming.dsc*), phishing (*www.seguridad.unam.mx/usuario-casero/phishing.dsc.*) and tips to prevent privacy and security breaches (www.seguridad.unam.mx/doc?ap=articulo&id=118) are provided to users.

United Kingdom

In the United Kingdom, the Home Office Identity Fraud Steering Committee launched the www.identity-theft.org.uk website which also contains tips about how to avoid ID theft. In addition, the Information Commissioner's Office has produced educational materials on ID theft through an information toolkit; television advertisements; and a training DVD.

United States

In May 2006, the US Federal Trade Commission launched "Deter, Detect, Defend," a nationwide education campaign aimed at helping consumers take steps to reduce risks of ID theft; monitor their personal information; and quickly react when ID theft is suspected. As part of the campaign, the *ID Theft Consumer Education Kit*, helps organisations and communities inform consumers about how to reduce risks of ID theft and respond if it strikes. The kit includes:

- A booklet which provides step-by-step instruction and tools to aid in consumer education.
- A brochure.
- A 10-minute video DVD – featuring stories of how real ID theft victims responded.
- A CD-ROM containing all educational materials for easy reproduction; and
- An in-depth guidebook for ID theft victims.

In April 2007, the US President's Task Force on Identity Theft issued a report setting forth a strategic plan for addressing the challenges presented by ID theft (US FTC and US DOJ, 2007). One focus of the strategic plan is to educate stakeholders about keeping sensitive consumer data out of the hands of ID thieves. The strategic plan recommends a multi-year public education campaign by federal, state, and local authorities. The US also established a website for information about the task force, the report, and victim's rights – *www.idtheft.gov*.

For more detail, see Annex 4.2.

Co-ordination of education initiatives

Australia

In Australia and New Zealand, the Australasian Consumer Fraud Taskforce supports a co-ordinated approach to awareness raising and education initiatives. The Taskforce was formed in March 2005 and is a group of 18 government regulatory agencies and departments with responsibility for consumer protection regarding frauds and scams. The Taskforce also has a range of community, non-government and private sector organisations as partners in the effort to increase the level of scam awareness in the community.

The purpose of the Taskforce is for the government members to work together to:

- Enhance the Australian and New Zealand governments' enforcement activity against frauds and scams.
- Run an annual co-ordinated information campaign for consumers: the 'Scams Awareness Month' in February or March (timed to coincide with Global Consumer Fraud Prevention Month).
- Involve the private sector in the information campaign and encourage them to share information they may have on scams and fraud.
- Generate greater interest in research on consumer frauds and scams.

Belgium

In Belgium, the Federal Public Service Economy, Small and Medium Size Enterprises, Self-employed and Energy, the Federal Computer Crime Unit of the Federal Police, and the Centre for Investigation and Information of Consumer Associations (CRIOC) organise several information campaigns targeting, among others, ID theft. For example, the *Fraud Prevention Campaign 2006 "Arnaqué, moi? Jamais!,"* is a campaign organised within the framework of ICPEN, with a specific focus on ID theft, consumer fraud in telephone services, consumer fraud on the Internet. Information is distributed through: leaflets sent by mail or available in social services of major cities and the information shop of the Federal Public Service Economy. It is also available on the website of the Federal Public Service Economy(*http://mineco.fgov.be/protection_consumer/fraud_prevention/home_fr_001.htm*); the radio; press conference and publication in newsletters of external partners and newspapers; headers of credit card and telephone bills. The campaign is funded by the Federal Public Service Economy and is carried out with the help of external partners (Belgacom, National Lottery, Proximus, Mobistar, Base, Diners Club, Citibank, American Express, Europabank, Les Maisons de Justice, *etc.*).

Mexico

In Mexico, the eCrime working group formed by public and private entities including the Mexican Banks Association ("ABM"), the Industry Transformation National Chamber ("Canacintra"), the National Bank of Mexico ("Banamex"), the Bancomer Bank, the Mexican Internet Association ("AMIPCI"), the Preventive Federal Police, Federal Communications Commission ("COFETEL"), the National Bank and Values Commission ("CNBV"), Nic Mexico and the public University of Mexico ("UNAM"), was created to gather data on phishing trends and to cancel domain names associated with identity fraud.

United States

In the United States for example, United States Attorney's Offices participate in ongoing training seminars, and several law enforcement agencies – including the US Department of Justice, the US Secret Service, the US FTC, and the Federal Bureau of Investigation – along with the American Association of Motor Vehicle Administrators have jointly sponsored over 20 regional, one-day training seminars on ID theft for state and local law enforcement agencies across the country (US FTC, 2007a, Vol. II, p. 71-73).

Annex 4.2 United States Initiatives to Combat Identity Theft

Consumers' liability for unauthorised credit card charges

In testimony before the Ohio state legislature, the US FTC explained how the loss is allocated between individuals and businesses, stating that: [US] [federal law limits consumers' liability for unauthorised credit card charges to USD 50 per card as long as the credit card company is notified within 60 days of the unauthorised charge. See 12 C.F.R. § 226.12(b). Many credit card companies do not require consumers to pay the USD 50 and will not hold the consumers liable for unauthorised charges, no matter how much time has elapsed since the discovery of the loss or theft of the card. Consumers' liability for unauthorised debit card charges is limited to USD 50 in cases where the loss is reported within two business days, and to USD 500 if reported thereafter. See 15 U.S.C. § 1693g (a). In addition, if consumers do not report unauthorised use when they see it on their bank statement within 60 days of receiving the notice, they may be subject to unlimited liability for losses that occurred after that period. ID Public Entities, Personal Information, and Identity Theft, Hearing Before the Ohio Privacy and Public Records Access Study Comm. of the Ohio Senate and House of Representatives (2007) (prepared statement of the US FTC, delivered by Betsy Broder, Assistant Director of the Division of Privacy and Identity Protection).

Public consumer and user education and awareness

The US DOJ has supported several different initiatives, such as the Ohio Identity Theft Verification Passport program (US IDTTF, 2007). The US Department of Treasury has also developed an ID theft resource page and serves as the head of the Financial Literacy and Education Commission, which is also comprised of 19 other US federal agencies and bureaus. This Commission launched a website and toll-free hotline for financial literacy in 2004, *www.MyMoney.gov*. The Department of Treasury and the Federal Reserve Banks also sponsor *Go Direct*, a campaign to motivate people who receive federal benefit checks to use direct deposit. Direct deposit helps eliminate the risk of stolen checks and reduces fraud.

Empowering domestic enforcement bodies

The US Department of State's Bureau of Diplomatic Security is working on an initiative involving the prosecution of individuals who use the identities of deceased people to obtain US passports (US IDTTF, 2007, Vol. II., p. 52). In addition, the US Internal Revenue Service ("IRS") is targeting ID thieves involved in fraudulent tax refund schemes (US IDTTF, 2007, Vol. II., p. 51).

Other federal agencies in the United States have also increased their law enforcement efforts to fight ID theft. The US Secret Service is currently targeting suspects who are trafficking in government-issued documents to reduce the online sale and distribution of stolen personal and financial information (US IDTTF, 2007, Vol. II., p. 51).

Special enforcement initiatives have been conducted to combat ID theft. For example, the FBI Cyber Division has conducted Operation "Retailers & Law Enforcement Against Fraud" ("RELEAF"), an international investigative initiative directed at reshipping – the use of one or more people to receive merchandise that criminals fraudulently purchased with others' credit cards to evade detection – and money laundering. This initiative, which involves more than 100 private sector participants and numerous law enforcement agencies, has resulted in more than 150 investigations.

The US Immigration and Customs Enforcement has established Document and Benefit Fraud Task Forces in 11 US cities to enhance interagency communications and improve federal, state, and local agencies' efforts to combat fraud that, by its nature, encompasses ID theft. Other federal agencies also actively participate in ID theft task forces, including the Social Security Administration, the IRS Criminal Investigation Division, and the Department of State's Bureau of Diplomatic Security (US IDTTF, 2007, Vol. II., p. 65-66).

The individual and co-ordinated efforts of federal law enforcement agencies in the United States have resulted in the prosecution of thousands of ID thieves. The US Sentencing Commission, which drafts criminal sentencing guidelines for federal offenses, maintains data that show that more than 1 000 offenders have been sentenced for convictions under the ID theft statute, 18 U.S.C. § 1028(a)(7), since its enactment in October 1998. The number of sentenced cases under the statute has increased substantially every year, from 12 cases in Fiscal Year 1999 to 195 cases in Fiscal Year 2006. Between Fiscal Years 1999 and 2006, the average sentences for these ID theft cases have increased steadily from an average of 16 months of confinement to 25 months of confinement (US IDTTF, 2007, Vol. II., p. 47).

In addition, the FBI reports that as of 30 September 2006, it had 1 274 pending ID theft-related cases, and that in Fiscal Year 2006, it opened 493 ID theft-related cases. In Fiscal Year 2006, the USPIS opened 1 269 ID theft cases and made 1,647 arrests, the US Secret Service made 3 402 ID theft arrests, and the US Social Security Administration reported that it opened 1 482 cases involving misuse of social security numbers (US IDTTF, 2007, Vol. II., p. 50).

To complement this scheme, the United States has developed data collection and sharing mechanisms to facilitate law enforcement prosecutions of ID theft. As mentioned in Section IV, the ID Theft Data Clearinghouse of the FTC logs ID theft complaints. The database is used by the FBI, USPIS, and other law enforcement officials for investigations (US IDTTF, 2007, Vol. II., p. 55). The Internet Crime Complaint Center ("IC3"), which was created by the FBI and the National White Collar Crime Center, is another conduit for complaints about Internet-related fraud and ID theft. IC3 offers law enforcement and regulatory agencies a central repository for complaints related to Internet crimes and allows them to retrieve timely statistical data and assess current crime trends.

In addition, the USPIS uses the Financial Crimes Database, a national database available to all postal inspectors, to analyse mail theft and ID theft complaints received from various sources, including the financial industry (American Express, Discover, MasterCard, Visa); major mailers (Netflix, Blockbuster, GameFly); the Identity Theft Assistance Center ("ITAC") complaints; online mail theft complaints, USPIS field offices; USPIS' telephone complaints; and US Treasury Checks (US IDTTF, 2007, Vol. II, p. 56).

Data security breaches disclosure

In 2005, the US federal bank regulatory agencies also issued guidance for banks, savings associations, and credit unions relating to breach notification.[1] The guidance advises that the covered financial entities, as part of security programs implemented pursuant to the interagency security guidelines, should develop and implement a response program to respond to incidents of unauthorised access to or use of sensitive customer information. The response program must contain, among other things, procedures for notifying the primary federal regulator of the incident, notifying appropriate law enforcement authorities in situations involving federal criminal violations requiring immediate attention, and notifying customers when warranted. The guidance also establishes standards for financial institutions to notify customers in the event of unauthorised access or misuse of sensitive customer information. If the institution becomes aware of an incident of unauthorised access to sensitive customer information, it should conduct a reasonable investigation to determine the likelihood that information has or will be misused. If the institution determines that misuse of customer information has occurred or is reasonably possible, it should notify any affected customer as soon as possible. The notice should be given in a clear and conspicuous manner, and it should include a description of the incident, the type of customer information affected, the steps taken to protect the customers' information from further unauthorised access, a telephone number that customers can call for further information and assistance, and other information relevant to the incident.

In addition to the US FTC data security enforcement actions, the US federal bank regulatory agencies have also initiated several enforcement actions against institutions failing to implement adequate programs to safeguard customer information. For instance, the FDIC had 17 formal enforcement actions between 2002 and 2006 (US IDTTF, 2007, Vol. II, p. 12). In the past five years, the Federal Reserve Board has taken 14 formal enforcement actions, the Office of the Comptroller of Currency has initiated 18 formal actions, and the Office of Thrift Supervision has taken 8 formal enforcement actions. In addition, these federal bank regulatory agencies have also taken several enforcement actions against financial institution insiders who breached their duty of trust to customers, were engaged in ID theft-related activities, or were otherwise involved in serious breaches or the misuse of customer information. These enforcement actions have resulted in bans from working in the financial services industry, personal cease and desist orders restricting the use of customer information, the assessment of significant civil money penalties, and orders requiring restitution.

Private sector's role

Other US industry associations have offered guidance on improving information security. The National Association of Realtors has issued their own information security guidelines consolidating best practices of real estate agents, multiple listing services, and associations to improve their security safeguards (US IDTTF, 2007, Vol. II, p. 22). Major health care industry organisations such as the American Hospital Association and the American Medical Association produce informational materials such as handbooks and toolkits. In addition, these organisations partner with vendors to provide security and privacy guidance to their members and to enable compliance with HIPAA Security and Privacy Rules.

Note

1. 12 C.F.R. Part 30, Supp. A to App. B (national banks); 12 C.F.R. Part 208, Supp. A to App. D-2 and Part 255, Supp. A to App. F (state member banks and holding companies); 12 C.F.R. Part 364, Supp. A to App. B (state non-member banks); 12 C.F.R. Part 570, Supp. A to App. B (savings associations); 12 C.F.R. Part 748, App. B (credit unions).

Chapter 5. Private Sector Initiatives: What Role for Industry and Internet Service Providers?

A serious private-sector threat

According to a *Global Security Survey* conducted by Deloitte Touche Tohmatsu in 2006, 53% of responding financial institutions considered that phishing and pharming would be the most significant threat over the next 12 months, followed by 51% of respondents by viruses and spyware. Only 30% of them declared that they had implemented technological tools to prevent these types of attacks (DTT, 2006, p. 13).

However, in its *2006 Global information Security Study*, PricewaterhouseCoopers indicates that an increasing number of executives across all industries are taking the threat more and seriously, and they made incremental improvements in deploying security policies and technologies, as reflected by doubled security budgets (PwC, 2006).

ISPs (Internet service providers) have also deployed a range of initiatives to help customers protect their security on line. Many provide software to protect against viruses and spyware, as well as filter e-mails for spam and viruses before they reach the users in order to minimise the risk of malware being installed on machines. BT, in conjunction with Yahoo!, filters its customers' mailboxes to help prevent spam from reaching their inboxes.

E-mail providers like Yahoo! Mail, are also developing technologies, such as DomainKeys, offering a cryptography-based solution to help solve the problems of phishing and email forgery, since they validate the origin of e-mail messages. ISPs also follow many of the guidelines of the Anti-Phishing Working Group ("APWG") on how to prevent and mitigate the damage caused by phishing attacks. ISPs have finally merged with software providers and government bodies to create technical bodies, such as the Messaging Anti-Abuse Working Group, to ensure that Internet users are safer on line.

What is happening in the US

In the United States, the payment card segment of the financial services industry has adopted one set of data security standards, the Payment Card Industry Security Standards ("PCI Standards"), for all merchants and service providers that store, process, or transmit cardholder data. The PCI Standards are designed to ensure that merchants and service providers properly handle and protect cardholder account and transaction information. Visa has its own program, the Cardholder Information Security Program, to ensure date security compliance in accordance with PCI Standards.[1]

In the event of a data breach, organisations such as the American Bankers Association, the Financial Services Roundtable, and the Payment Card Industry have developed guidelines to address breach response issues. They encourage businesses to create an internal response plan that, among other things, confirms, analyses, and documents events and, in the event of a breach, they advise entities to immediately contain the breach and limit possible exposure of consumer information while preserving logs and other electronic evidence.[2]

Organisations, such as the Council of Better Business Bureaus and the National Cyber Security Alliance, have produced primers for small business on information including: the importance of employee screening and training in the use of physical safeguards beyond electronic measures to prevent identity theft, tips for recognizing attempts at ID theft, and guidance on the handling and management of sensitive information.

Other businesses have adopted industry and government guidelines on how to detect and avoid malware, including guidelines developed by the US National Institute of Standards and Technology ("NIST") (US IDTTF, 2007, Vol. II, 25). NIST's recommendations for improving an organisation's malware incident prevention measures include: planning and implementing an approach based on the most likely point of attack, ensuring that policies address the prevention of malware incidents and include provisions relating to remote workers, and using appropriate techniques to prevent malware incidents (*e.g.* patch management, application of security configuration guides). In addition, some private sector entities have developed standards and guidelines to ensure that third party service providers, such as outsourced IT operations, adhere to the contracting parties' security requirements.

Despite the substantial effort undertaken by public and private sector entities on how to respond to data breaches, surveys of large corporations and retailers in the US indicate that fewer than half of them have formal breach response plans (US IDTTF, 2007, Vol. I, p. 35).

For example, an April 2006 cross-industry survey showed that only 45% of large multinational corporations headquartered in the United States had a formal process for handling security violations and data breaches. Fourteen percent of the companies surveyed had experienced a significant privacy breach in the past three years. A July 2005 survey of large North American corporations found that although 80% of responding companies reported having privacy or data protection strategies, only 31% had a formal breach notification procedure. Moreover, one survey found that only 43% of retailers had formal incident response plans, and an even smaller percentage of retailers had actually tested their plans.

In Mexico, the eCrime working group formed by public and private entities, including the Mexican Banks Association ("ABM"), the Industry Transformation National Chamber ("Canacintra"), the National Bank of Mexico ("Banamex"), the Bancomer Bank, the Mexican Internet Association ("AMIPCI"), the Preventive Federal Police, Federal Communications Commission ("COFETEL"), the National Bank and Values Commission ("CNBV"), Nic Mexico and the public University of Mexico ("UNAM"), was created to gather data on phishing trends and to cancel domain names associated with identity fraud.

Private sector consumer-education tools

Financial institutions, which can be primary victims in a phishing attack, increasingly alert their customers about new phishing messages and security risks. In the United States, since 2004, financial institutions have undertaken joint educational efforts through the ITAC, a domestic organisation representing some of the largest US banks, brokerages, and finance companies.[3] In addition, the NASD issued an alert entitled "Phishing and Other Online Identity Theft Scams: Don't Take the Bait" (US IDTTF, 2007, Vol. II, p. 40).

The National Cyber Security Alliance maintains *Stay Safe Online*, a website with safe computing tips such as how to stop suspect e-mails and websites.[4] Similar outreach efforts have been made by other private organisations. For example, Microsoft and Best Buy, in conjunction with other private and public partners, sponsor the Get Net Safe Tour which sends experts to schools, Internet fairs, and community centers to discuss Internet safety (US IDTTF, 2007, Vol. II, p. 39).

Many US colleges and universities have also tried to increase awareness of ID theft among vulnerable college students through websites, orientation campaigns, and seminars. The National Council of Higher Education Loan Programs (NCHELP) reached out to its constituents and encouraged them to take advantage of the US FTC's resources on preventing ID theft and share them with students (US IDTTF, 2007, Vol. II, p. 42).

In the United Kingdom, various banking and payment associations such as the Financial Services Authority ("FSA"), the British Banker Association ("BBA"), and the UK payments Association ("APACS"),[5] have pro-actively developed initiatives to educate both their own members (*i.e.* banks, and companies) and their customers (OFCOM, 2006, p.37). Even in countries where ID theft is not yet perceived as a serious threat, financial institutions alert their customers about how to keep safe on line. In the Netherlands for example, *Nederlands Vereniging van Banken* ("NVB"), the Dutch Banking Association, began an awareness campaign in 2006 informing consumers about the risks of ID theft and how to better protect their personal information (INTERVICT, 2006, p. 24).

In Mexico, some members of the Internet Mexican Association (AMIPCI) have created a website *www.navegaprotegido.com.mx*, which contains information aimed at educating consumers on the risks of ID theft.

Some security vendors have also put in place alert systems to inform their customers about online risks. For instance, under its *Security Connection* service, Symantec, in conjunction with financial institutions, provides customers of these financial institutions with tools to make them aware about ongoing online threats and to enable them to recognise, avoid, and denounce them. Typically, customers accessing financial institutions' websites are routed to Symantec *Security Connection* where articles, expert advice, and real time security alerts are available.

In addition, ISPs have developed educational initiatives around the risks and precautions users should take when operating on line. For instance, the UK Yahoo!'s Security Centre provides information on the latest virus alerts as well as tips on how users can stay safe on line.

Legal right of action

Few countries provide ISPs with a right of action, on behalf of their customers, against perpetrators of "phishing" attacks.[6] In most countries, in a successful phishing scam, only the affected Internet user is entitled to take legal action against the phisher, to the exclusion of ISPs or the spoofed company. In France for example, when the law on the digital economy was debated in the French Parliament, the national Association of Internet Service Providers (*Association des Fournisseurs d'Accès et de Services Internet* (AFA))[7] proposed that such a right of action be extended to ISPs on behalf of their customers. This proposal was going in the same direction as one recommended measure in the *OECD 2006 Anti-Spam Toolkit* which provides that "ISPs should be able to take appropriate and balanced defensive measures to protect their networks, and should be allowed to take legal action against spammers" (OECD, 2006c, p. 12). Although AFA's proposal was eventually rejected by the French Parliament, the Association continues to request French authorities to consider it.

Noting that in a typical phishing scam each victim may not sustain losses significant enough, or have the resources to pursue enforcement action individually, Microsoft has recently re-advocated that ISPs and aggrieved businesses should be allowed, in the EU, to take legal action against cybercriminals (Microsoft, 2006) to protect their customers and recoup their own significant losses by bringing damage claims for the costs they incur from cybercrime, including reputation and consequential damages.

Notes

1. See: http://use.visa.com/business/accepting_visa/ops_risk_managment/cisp.html.
2. American Express, Data Compromise Workbook (2006), at 6-8 (quoted in US IDTTF, 2007, Vol. II, p. 27).
3. ITACs' actions may be consulted at: *www.identitytheftassistance.org/.*
4. See: www.staysafeonline.org/basics/consumers.html.
5. APACS maintains www.banksafeonline.org.uk/.
6. In 2006, based on the US state of Virginia's anti-phishing law, the US *Federal Lanham Act*, and the US *Federal Computer Fraud & Abuse Act*, AOL filed three civil lawsuits against unnamed groups who targeted AOL and CompuServe members with phishing scams.
7. See: *www.afa-france.com/t_spam.html.*

Annex 5.1 Private-Sector Initiatives to Educate Consumers about ID Theft

In some countries, the private sector (in addition to the government) has supported education initiatives on ID theft. Below are some examples:

Australia

In Australia, the Australian Bankers' Association (ABA), the Australian High Tech Crime Centre and the Australian Securities and Investments Commission (ASIC), jointly support a website, *Protect Your Financial Identity* (www.protectfinancialid.org.au), which assists people in protecting their financial identity and minimising the damage if a problem occurs. The website contains practical prevention tips, fact sheets, and an interactive quiz allowing people to test how secure their personal details are.

Mexico

In Mexico, some members of the Internet Mexican Association (AMIPCI) have created a website www.navegaprotegido.com.mx, which contains information aimed at educating consumers on the risks of ID theft.

The Netherlands

In the Netherlands, *Nederlands Vereniging van Banken*, the Dutch Banking Association, began an awareness campaign in 2006 informing consumers about ID theft risks and how to protect their personal information (INTERVICT, 2006, p. 24).

United Kingdom

In the United Kingdom, various banking and payment associations, such as the British Banker Association (BBA) and the UK payments Association

(APACS), have pro-actively developed initiatives to educate both their own members (*i.e.* banks, and companies) and their customers; further information is available on the Internet at *www.banksafeonline.org.uk* (OFCOM, 2006, p. 37).

United States

In the United States, several different industries are active in educational initiatives to help fight ID theft. For example, financial institutions, which can be primary victims in a phishing attack, increasingly alert their customers about new phishing messages and security risks. Since 2004, financial institutions have undertaken joint educational efforts through the Identity Theft Assistance Center, a domestic organisation representing some of the largest US banks, brokerages, and finance companies.

In addition, the National Association of Securities Dealers has published a guide entitled "Phishing and Other Online Identity Theft Scams: Don't Take the Bait." More recently, the Identity Theft Prevention and Identity Management Standards Panel ("IDSP") sponsored by the US Better Business Bureau (BBB) and the American National Standards Institute ("ANSI") created a new market-wide initiative that would help arm businesses and other organisations with the tools they need to combat ID theft and fraud and protect consumers – and themselves – from the risks associated with these crimes (http://www.ansi.org/standards_activities/standards_boards_panels/idsp/report_webinar08.aspx?menuid=3). The report contains a catalogue of existing standards, best practices and related compliance systems germane to this issue across all market sectors and industries as well as recommendations for areas where the government and private sector should develop additional standards and guidelines. The report contains specific recommendations for consumer education initiatives in areas such as security freezes.

Chapter 6. International, Bilateral and Regional Initiatives

Various international forums – including international organisations and informal networks focusing on security, privacy or consumer protection issues – devote substantive efforts to tackling cybercrime's multiple facets. This chapter highlights these organisations and their efforts.

International organisations

The OECD

The OECD has been involved with efforts to build user trust on line through more effective cross-border law enforcement. It has analysed the challenges and developed policy responses in the areas of fraud against consumers, spam, security and privacy, through the following main instruments:

- E-commerce Guidelines (OECD, 1999): these Guidelines reflect existing legal protection available to consumers in more traditional forms of commerce; encourage private sector initiatives that include participation by consumer representatives; and emphasise the need for co-operation among governments, businesses and consumers. Their aim is to encourage: fair business, advertising and marketing practices; clear information about an online business' identity, the goods or services it offers and the terms and conditions of any transaction; a transparent process for the confirmation of transactions; secure payment mechanisms; fair, timely and affordable dispute resolution and redress; privacy protection; and consumer and business education.
- Cross-border Fraud Guidelines (OECD, 2003): these Guidelines establish a common framework to combat online and offline cross-border fraud through closer, faster, and more efficient co-operation between consumer protection enforcement agencies. As concluded in the OECD report on the implementation of the Guidelines, (OECD, 2006c), within three years, OECD member country domestic and international co-operation systems have been improved through:

strengthened enforcement agencies; enhanced national co-ordination between government bodies and with the private sector; intensive consumer education about the challenges of cross-border fraud; and the exchange of information and best practices at the international level.

- Recommendation on Consumer Dispute Resolution and Redress (OECD, 2007a): the Recommendation addresses the current practical and legal obstacles to pursuing remedies in consumer cases, whether locally or in a cross-border context.
- Security Guidelines of Information Systems and Networks (OECD, 2002): these Guidelines provide a framework of principles to foster consistent domestic approaches in addressing security risks in a globally interconnected society. Following their release, the OECD monitored efforts by governments to implement national policy frameworks, including protecting critical information infrastructures, combating cybercrime, developing Computer Security Incident Response Teams, raising awareness, and fostering education (OECD, 2007d). In 2006, the OECD issued a Guidance on Electronic Authentication setting out a number of operational principles aimed at helping member countries establishing or modernising their approaches to authentication. This guidance was followed in 2007 by the adoption of an OECD Recommendation on Electronic Authentication (OECD, 2007c), encouraging member countries to establish compatible, technology-neutral approaches for effective domestic and cross-border electronic authentication of persons and entities.
- Guidelines on the Protection of Privacy and Transborder Flows of Personal Data (OECD, 1980): these Guidelines set out core principles aimed at assisting governments, business and consumer representatives in their efforts to protect privacy and personal data, and in obviating unnecessary restrictions to transborder data flows, both on and off-line. In October 2006, the OECD published a report describing member country current attempts to address privacy challenges. Despite these efforts, the report highlights the need for a more global and systematic approach to cross-border privacy law enforcement co-operation (OECD, 2006a). Responding to this conclusion, in 2007, the OECD adopted a Recommendation on Cross-border Co-operation in the Enforcement of Laws Protecting Privacy (OECD, 2007b), which sets out principles aimed at enabling member country authorities to co-operate with foreign authorities, as well as to provide mutual assistance to one another in the enforcement of privacy laws.
- Anti-Spam Toolkit (OECD, 2006b): Aimed at facilitating international co-operation in the fight against spam, the toolkit is based on the

premise that various elements need to be brought together to bear on the problem of spam and help develop anti-spam strategies and solutions in the technical, regulatory and enforcement fields. It provides a set of recommendations aimed at establishing consistent and complementary policies and other (e.g. enforcement) anti-spam initiatives among OECD member countries.

The OECD also co-operates with other international forums to ensure effective cross-border co-operation. The Asia-Pacific Economic Co-operation ("APEC")[1] and the OECD have joined forces to provide policy makers in their economies with a global picture of the malware phenomenon. This collaboration is undertaken under a joint project conducted with the help of an "APEC-OECD group of experts," composed of delegates to both groups, representatives of Computer Emergency Response Teams ("CERTs"),[2] and enforcement bodies. In March 2007, the OECD and APEC held a workshop on malware. As mentioned earlier,[3] this event informed a joint analytical report aimed at helping governments identify priority areas of intervention and provide policy orientation to develop appropriate strategies to combat the proliferation of malware (OECD, 2008).

APEC member economies have undertaken specific strategies to establish regulatory frameworks providing the essential underpinnings for growth and consumer confidence in the digital marketplace. In 2002, APEC endorsed its *Cyber Security Strategy* recommending its member economies to adopt legislation and policies criminalising cybercrime. In 2005, it further urged its members to address the threat posed by the misuse, malicious and criminal use of the online environment under its *Strategy to Ensure a Trusted, Secure and Sustainable Online Environment* (APEC, 2005).

The International Telecommunication Union ("ITU") and the Asia-Europe meeting ("ASEM")[4] are other platforms where policy makers and law enforcers around the globe explore and recommend policy actions that could be adopted to improve international co-operation against cyber attacks.

The United Nations

In 2005, an open-ended Intergovernmental Expert Group To Prepare a Study on Fraud and the Criminal Misuse and Falsification of Identity ("UN IEG") was set up to address the problem of fraud and criminal misuse of identity. In 2007, it issued a report containing the results of a 2005-2006 study,[5] commissioned by the United Nations Office on Drugs and Crimes ("UNODC"), on fraud and the criminal misuse and falsification of identity. The report, which integrates comments from notably the UNODC and the

United Nations Commission on International Trade Law ("UNCITRAL"), contains a number of recommendations for best practices to be implemented by governments and the private sector (UN IEG, 2007, pp. 5-20 and see Annex 6.2 at the end of this chapter).

As recommended in the UN Commission on Crime Prevention and Criminal Justice's 2007 *Resolution on International Co-operation in the Prevention, Investigation, Prosecution and Punishment of Economic Fraud and Identity-related Crime*, the UNODC set up a consultative platform of experts from governments, international organisations (including the OECD), and the private-sector, to pool experience and develop strategies on identity-related crime. The first meeting of the expert group was held in Italy in November 2007, in conjunction with an International Conference on identity-related crime.[6] The UNODC will build on the conclusions of both the expert group meeting and the International Conference to develop guidelines, best practices and training materials in the prevention, investigation and prosecution of identity-related crime.

Interpol

Interpol is the international police organisation whose mission is to prevent or combat international crime.[7] Interpol frequently serves as the basis of co-operation between national police forces in conducting multinational investigations of online crime. Interpol has decentralised its cyber crime expert teams around the world through the establishment of regional Working Parties on Information Technology Crime for Europe, Latin America, Asia, South Pacific and Africa.[8] Interpol's European Working Party on Information Technology Crime ("EWPITC") has, for example, compiled a best practise guide for experienced investigators from law enforcement agencies.[9] It has also set up a rapid information exchange system under an international 24-hour response scheme, listing responsible experts within more than 100 countries. This scheme was notably endorsed by the G8 24/7 High Tech Crime Network ("G8 24/7 HTCN"). The EWPITC further agreed to conduct a project aimed at enabling law enforcement agencies worldwide to investigate botnets and malicious code. This work will facilitate an operational co-operation project of European BotNet investigators to share intelligence and best practices.

International informal networks

Policy frameworks are complemented by informal international networks of enforcement agencies fighting against fraud and spam.

- International Consumer Protection Enforcement Network ("ICPEN")

ICPEN is an informal network where domestic enforcement authorities join forces to enhance international consumer protection enforcement. It includes authorities from 36 countries plus two observers (the OECD, and the European Commission). ICPEN is mainly used as a platform of collaboration where member country enforcement bodies exchange information on fraud cases affecting consumers from their jurisdiction through monthly teleconferences, national reports, and the *econsumer.gov* website.

- London Action Plan ("LAP")

The LAP is a global co-operation network of governments and private sector representatives from around the world focused on combating spam and increasing international anti-spam co-operation. The plan was developed by the US FTC and the UK OFT in 2004 to curb the activities of international spammers, and now includes participants from more than 20 countries, including more than 30 government agencies and 20 private sector representatives. Key elements of the LAP include designating a point of contact for further enforcement communications, exchanging effective investigative techniques and enforcement strategies, and discussing consumer/business education. In November 2006, in conjunction with the Contact Network of Spam Enforcement Authorities ("CNSA"),[10] the LAP held a joint workshop on international enforcement co-operation against spam, spyware and other online threats. In October 2007, the LAP held another joint meeting with the CNSA that also featured joint sessions with the Messaging Anti-Abuse Working Group ("MAAWG").

- G8 24/7 High Tech Crime Network ("HTCN")

The G8 24/7 HTCN is another informal network which provides around the clock high-tech expert contact points[11] which permits the sharing of information on ongoing investigations against cyber criminals. Created in 1997, the G8 24/7 HTCN, which includes 45 countries, has, among other achievements, been used on several occasions to avert hacking attacks, including attacks on banks in the United States, Germany and Mexico.[12] In its report on fraud and the criminal misuse and falsification of identity,[13] the UN IEG recommends that Member States make use of the G8 24/7 HTCN, both for emergency and non-emergency cases involving electronic fraud or identity-related crimes (UN IEG, 2007, paragraph 27, *d*).

Bilateral and regional law enforcement frameworks

Bilateral efforts

- Canada - United States Cross-Border Crime Forum[14]

This bi-national working group provides governments of the two countries with reports and recommendations laying the groundwork for substantial improvements in enforcement capabilities, bi-national co-ordination and co-operation in combating trans-national fraud. The working group is expected to assist the US Identity Theft Task Force to co-ordinate education, prevention and enforcement in relation to phishing, and ID theft. In October 2006, the Forum's Working Group on Cross-Border Mass-Marketing Fraud published a report on phishing and its impact on cross-border crime (BWGCBMMF, 2006). The report refers to several enforcement actions between Canadian and US authorities against phishers.

- Australasian Consumer Fraud Task Force ("ACFTF")

The ACFTF[15] is a group of 18 government regulatory agencies and departments from Australia and New Zealand responsible for consumer protection regarding frauds and scams. The Task Force runs the Consumer Awareness Month ("CAM"), a campaign educating consumers about scams. During the March 2006 CAM, the ACFTF focused on phishing scams, noting the lack of awareness among Australians who seem to be responding to these scams at a much higher rate than other countries.[16]

- The Police Commissioners' Conference Australasian Identity Crime Working Party

To implement the Australasian Identity Crime Policing Strategy 2003-2005, a Police Commissioners' Conference Australasian Identity Crime Working Party was established where police representatives from all Australian jurisdictions and New Zealand, the Australian Crime Commission, and the Australasian Centre for Policing Research develop a work plan to improve their fight against identity crime. Under its 2006-2008 Strategy (ASWPRPCC, 2005), the Working Party recommended the development of measures including: the allocation of appropriate resources by law enforcement to combat identity crime including; the provision of appropriate training to allow the implementation of best practice policing techniques in all jurisdictions; effective partnerships between local, Australasian, regional and international law enforcement authorities and with policing and public and private sector organisations.

Regional efforts

- The Council of Europe Convention on Cybercrime[17]

The Convention is the first and only legally binding multilateral treaty addressing the problems posed by the spread of criminal activity on line. Signed in Budapest in 2001, the Convention entered into force on 1 July 2004. Taking account of digitalisation, convergence and continuing globalisation of computer networks, the Convention requires its Parties to establish laws which criminalise security breaches resulting from hacking, illegal data interception, and system interferences that compromise network integrity and availability.

This instrument, whose Preamble notably cites OECD actions as a means to further advance international understanding and co-operation in combating cybercrime, aims at "pursuing … a common criminal policy for the protection of society against cybercrime by adopting appropriate legislation and fostering international co-operation." To achieve these goals, the Parties commit to establish certain substantive offences in their laws which apply to computer crime. Although online identity theft is not *per se* mentioned in the Convention among the illegal activity that signatories must criminalise, it is indirectly covered under closely related crimes including illegal access to computers, illegally accessing computer data, and computer-related fraud that are listed in the Treaty.[18]

Parties to the Convention also agree to adopt domestic procedural laws to ensure that their law enforcement bodies have the necessary authority to deter, investigate and prosecute cybercrime offences and actively participate in international co-operation efforts. Such involvement should take the form of mutual assistance[19] for the purposes of investigations or proceedings concerning criminal offences related to computer systems and data, or for the real-time collection of traffic data that could evidence a criminal offence. Measures for the preservation of data are also set out. Moreover, the Convention invites its Parties to spontaneously share with each other information obtained within the framework of their own investigations when they consider that the disclosure of such information might assist another party in initiating or carrying out investigations or proceedings.[20]

As such, the Convention encourages a more coherent approach in the fight against cyber attacks. For example, it has been recognised as an important international benchmark for the development of cybercrime laws by APEC countries (APEC, 2006). Some companies in the private sector have taken some initiatives to help ensure a larger impact of the Convention's principles (Microsoft, 2006).[21] However, to date, the Convention has been mostly ratified by a majority of EU Member States and its ratification by many countries outside the EU is still needed to allow for a larger-scale implementation.[22]

- European Union (EU) instruments
- The EU e-communications regulatory framework

Several EU Directives set out measures to fight cyber attacks. Among them, Directive 95/46 on the Protection of Individuals with Regard to the Processing of Personal Data[23] lays out conditions for legitimate data processing (whether offline or online), one of which being that personal data may be only processed if the data subject has unambiguously given his consent.[24] The e-privacy Directive[25] reiterates this condition. Considering that users are very often not aware of the fact that outsiders can gain access to their computers and store information or programs with no means to control such illicit and hidden activity, Article 5.3 of the Directive states that "access to information stored in the terminal equipment of a subscriber or user is only allowed on condition that [he] is provided with clear information about the purposes of the processing and is offered the right to refuse such processing by the data controller." This provision does not only apply to so-called spyware[26] (hidden espionage programs) and Trojan horses (programs hidden in messages or in other innocent looking programs) but also to cookies (tracking devices which register users' preferences as they visit websites).[27] In addition, Article 13 of the Directive imposes a ban on spam.

Taking a significant step further, to improve co-operation between law enforcement, judicial authorities and the police in their fight against attacks on information systems, the EU Council, in its 2005/222/JHA Framework Decision,[28] called for an EU-wide approximation of criminal laws in this area. Considering that the measures adopted to date by EU Member States to introduce effective, proportionate and dissuasive sanctions to combat attacks against information systems were not sufficient, the Council, on the basis of the EC Treaty principle of subsidiarity,[29] imposed such a harmonisation on EU Member States.

Under Article 2 of the Framework Decision, each EU Member State will have to take the necessary measures to ensure that the intentional access to the information system, illegal system and data interference are punishable as a criminal offence. Article 6 of the same decision further states that illegal system and data interference shall be punishable by criminal penalties of at least between one and three years of imprisonment. All these measures were to be adopted and implemented into EU Member States' legislation by 16 March 2007.

In November 2006 the European Commission adopted a Communication on *Fighting Spam, Spyware and Malicious Software*.[30] This Communication takes stock of the efforts made so far to fight these practices and concludes that there is currently insufficient action to address these threats to the Information Society. The Communication calls upon Member States to step up enforcement efforts using the already existing instruments and involve market players drawing on their expertise and available knowledge.

- EU-specific actions in relation to ID theft

Although no existing EU legislation specifically tackles ID theft, the EU has recently recognised the need for specific initiatives, proposing to consider it as an offence *per se*. The European Commission indeed sees the problem of ID theft as clearly growing, noting that the percentage of global credit card fraud which occurs in Europe grew from 12% in 2002 to 17% in 2003 (European Commission, 2004).

The European Commission's 2004-2007 Action Plan on the prevention of fraud on non-cash means of payment suggested examining *i*) the merits of establishing an EU single contact point for individuals against ID theft which could include a register of bodies engaged in the prevention of this illegal activity; *ii*) the creation of a database of original and counterfeit identity documents accessible to public authorities and the private sector. The Commission also reinforced the role of the EC Fraud Prevention Expert Group ("FPEG"), an independent body, chaired by the Commission, whose main objective is to intensify co-operation between interested parties, especially at the international level.[31] Various law enforcers, banks, retailers, consumer groups and network operators are members of the EU FPEG and exchange information and best practice to prevent fraud. In October 2007, its subgroup on ID theft issued a *Report on Identity Theft/Fraud* in the financial sector (EC FPEG, 2007), with a particular focus on payment and retail banking areas. The paper notably contains a number of recommendations for more tailored and consistent policy measures across the EU.

In parallel, the Commission's Directorate General for Justice, Freedom and Security ("DG FJS") organised in November 2006 a high level conference on the integrity of identities and payments to further strengthen public-private co-operation against identity theft and credit cards fraud.[32] In May 2007, the Commission adopted a Communication on a general policy on the fight against cybercrime announcing a series of actions including the launching of an in-depth analysis aimed at preparing a proposal for specific EU legislation against identity theft (EC, 2007, p. 10). On that basis, in July 2007, the Commission's DG FJS launched a comparative study on ID theft, examining the various definitions used in EU Member States and exploring the idea of criminalising the offence across the EU.

- The EU Regulation on Consumer Protection Co-operation ("CPC Regulation")33

Responding to the development of the Euro, e-commerce, and EU enlargement, the 2004 CPC Regulation calls on EU Member States to institute a minimum level of common investigation, enforcement, and co-operation powers among consumer protection enforcement bodies. The new Regulation, which entered into force on 1 January 2007 in all EU and EEA Member States, puts in place mechanisms for effective enforcement co-operation against consumer fraud. It is notably aimed at tackling online scams.

Under Article 6.1 of the Regulation, a consumer protection agency shall, upon request from an applicant authority, "supply without delay any relevant information required to establish whether an intra- Community infringement has occurred or to establish there is a reasonable suspicion it may occur." It may also do so without request when it becomes aware of an infringement or where it suspects that such an infringement may occur. Under Article 8.1 of the Regulation, authorities in a Member State have the obligation, on request from their counterparts in other Member States, to take all necessary enforcement measures to bring about the cessation of an infringement without delay. The new framework thereby eliminates a situation where a public authority is prevented, under its national confidentiality rules, from communicating the necessary information requesting assistance from an authority in the member state of the trader. The Commission is creating a database through which consumer protection enforcement bodies will be able both to communicate with each other and to notify the existence of any unlawful intra-Community practices.

- EU agencies and bodies

Several agencies and bodies assist EU institutions in their fight against cyber crime. Among them, the following bodies are particularly active in the fight against identity theft:

  - *Europol* provides expertise and technical support for investigations and operations related to cyber crime. In late October 2004, it participated in "Operation Firewall," an international sting to bring down a phishing operation. On that occasion, Europol co-ordinated efforts with the US *Secret Service*, the *Department of Justice* and *Homeland Security Department*, the *Royal Canadian Mounted Police*, the Vancouver Police Department's Financial Crimes Section, the

UK National Hi-Tech Crimes Unit, and local police from *Belarus*, the Netherlands, *Sweden* and *Ukraine* to arrest *28 individuals from* 8 states and 6 countries for using websites to steal identities, credit cards, and identity documents. This criminal organisation was successful in stealing one million credit card numbers and in creating fraudulent documents such as birth certificates.34

- The EU Contact Network of Spam Enforcement Authorities ("CNSA"): CNSA is a forum which facilitates the sharing of information and best practices in enforcing anti-spam laws between the national authorities of EU Member States and of the EEA. In February 2005, CNSA signed a voluntary agreement establishing a common procedure for handling cross-border complaints on spam. As mentioned above, in November 2006, CNSA and the LAP held a joint workshop on international enforcement co-operation against spam, spyware and other online threats. Speakers concluded that links with other international players fighting spam should be made to ensure enhanced international co-operation in the enforcement of anti-spam laws as well as education and awareness campaigns. In October 2007, the LAP will hold another joint meeting with CNSA that will also feature joint sessions with the MAAWG.
- The European Network and Information Security Agency ("ENISA"): ENISA is an EU agency funded by the European Commission. In 2006, it conducted a survey on measures taken by Internet service providers to safeguard the security of their services and fight against spam, spyware and other forms of malware (ENISA, 2006).

Public-private international enforcement initiatives

The protection of the Internet infrastructure is a shared responsibility of business and government. Various international schemes where business joins public authorities' efforts to tackle cyber attacks come as a complement to government initiatives.

Data Sharing Forums

- The Anti-Phishing Working Group ("APWG")

The APWG[35] defines itself as a global pan-industrial and law enforcement association focused on eliminating fraud and identity theft that result from phishing, pharming and e-mail spoofing of all types. It publishes monthly reports containing statistics on phishing aiming at measuring the scale of the threat. [36] It is largely co-ordinated via e-mail communications and meets periodically. On 30-31 May 2007, the APWG held the *Counter-eCrime Operations Summit*, which engaged "questions of operational challenges and the development of common resources for the first responders and forensic artisans who confront the e-crime threat."

- DigitalPhishNet ("DPN")

DPN[37] is a collaborative forum where ISPs, online auction sites, financial institutions, and law enforcement agencies share statistics and best practice to tackle phishing and other online threats. DPN permits the sharing of information about phishers in real time to assist with identification, arrest, and prosecution.

- WHOIS

Law enforcement agencies and business38 typically use WHOIS services, a public directory of domain name information. When a domain name is registered, the owner of the domain name's postal address, e-mail address and phone number are automatically published in WHOIS. WHOIS also enables access to data on Internet Protocol address allocations that can identify the physical locations where unlawful activity is taking place, and the relevant ISPs, which, in turn, can provide information regarding their customers. The Internet Corporation for Assigned Names and Numbers ("ICANN"), the non-profit body responsible for accrediting domain name registrars, requires that this personal information be accurate and available for anybody to view on the Internet.

Consumer and industry enforcement strategies

While laws can be slow to adapt to technological advances, in parallel to their participation in law enforcers' activities, businesses have developed their own enforcement strategies.

- The Trans Atlantic Consumer Dialogue ("TACD")

TACD is a forum where US and EU consumer organisations develop and agree joint consumer policy recommendations addressed to the US government and EU to promote consumers' interest in EU and US policy making. On 20 February 2007, TACD issued a *Resolution on ID theft, phishing and consumer confidence*[39] recommending US and EU authorities to: adopt more specific laws prohibiting malware, spyware, ID theft and phishing; in particular, laws on ID theft should impose increased sanctions on fraudsters and should provide a duty for companies to disclose data security breach information to their customers. TACD also requests business to compensate victims for their financial and non-financial losses following ID theft (TACD, 2007).

- The Anti-Spyware Coalition ("ASC")

The ASC is a group composed of anti-spyware software companies, academics, and consumer groups. It focuses on the development of standard definition in relation to spyware. On 25 January 2007, the ASC published working documents on best practices[40] aimed at detailing the process by which anti-spyware companies identify software applications as spyware or other potentially unwanted technologies.

- The Messaging Anti-Abuse Working Group ("MAAWG")

MAAWG is an international organisation focused on preserving the electronic messaging from online exploits and abuse such as messaging spam, virus attacks, denial-of-service attacks, and other forms of abuse. These messaging service providers seek to enhance user trust and confidence.[41] On 15 May 2007, MAAWG issued the *MAAWG Sender Best Communications Practices* with collaborative input from both volume senders and Internet Service Providers. The *Best Practices guidelines* are a milestone toward industry agreement on how senders can distinguish their legitimate volume e-mail from unsolicited spam.[42]

- Private companies' actions (examples)

Some companies have implemented their own tools to combat cyber attacks. On 31 January 2007, Symantec launched its *Online Identity Initiative*, providing consumers with software enabling them to manage their identity safely while on line.[43]

Microsoft's *Anti-Virus Reward Program*[44] is another tool whereby the company will provide monetary rewards for information resulting in the arrest and conviction of parties launching malicious viruses and worms on line. In January 2006, the company provided investigative and technical support to Bulgarian authorities to help arrest eight members of an international criminal network whose phishing scam spoofing MSN's name invited consumers to reveal their personal information by "updating" their personal data such as credit card account details. The group had launched a co-ordinated attack on 43 servers located in 11 countries and defrauded US, German and UK credit card holders of more than USD 50 000.[45]

Notes

1 The 21 APEC Member Economies are as follows: Australia, Brunei Darussalam, Canada, Chile, People's Republic of China, Hong Kong China, Indonesia, Japan, Republic of Korea, Malaysia, Mexico, New Zealand, Papua New Guinea, Peru, Philippines, Russia, Singapore, Chinese Taipei, Thailand, United States, and Viet Nam.

2 CERTs are security alert centres present in all OECD member countries, involved in the prevention of computer attacks. Their alerts, which inform companies and/or administrations, are publicly accessible.

3 See Introduction in this book.

4 The Asia-Europe Meeting is a multilateral forum for action-orientated debate between the EU Member States and Asian partner countries including Brunei, Burma, Cambodia, China, Indonesia, Japan, Korea, Laos, Malaysia, the Philippines, Singapore, Thailand, and Vietnam. At its 4th Conference on E-commerce in 2005 in London, ASEM examined the question of e-commerce crimes.

5 The following Member States responded to the UN study: Algeria, Belarus, Canada, Costa Rica, Croatia, Egypt, Finland, Germany, Greece, Hungary, Italy, Japan, Jordan, Republic of Korea, Latvia, Lebanon, Former Yugoslav Republic of Macedonia, Madagascar, Malta, Mauritius, Mexico, Monaco, Morocco, Netherlands, Nicaragua, Norway, Oman, Panama, Peru, Romania, Russian Federation, Slovak Republic, Slovenia, South Africa, Spain, Sudan, Sweden, Switzerland, Syria, Trinidad and Tobago, Turkey, United Arab Emirates, United Kingdom, United States, Zambia.

6 The Conference was co-organised by the UNODC, the United Nations International Scientific and Professional Advisory Council ("ISPAC"), the *Centro Nazionale di Prevenzione e Difesa Sociale* ("CNPDS") and the Courmayeur Foundation.

7 Interpol now has 186 member countries. See: *www.interpol.int/public/icpo/default.asp*.

8. See: *www.interpol.int/Public/TechnologyCrime/WorkingParties/Default.asp#europa*

9. This best practice guide is called *The Information Technology Crime Investigation Manual*. It is digitally available via Interpol's restricted website.
10. See more details about CNSA in the section below on *Regional schemes*.
11. As notably recommended by APEC in its above mentioned *Cyber Security Strategy* and the UN IEG in its 2007 report.
12. See details about the G8 24/7 HTCN at: *www.networkworld.com/news/2006/102306dojofficialcybercrimecooperation.html?page=2*.
13. Reference to the UN report is made in the subsection on the United Nations of the present paper.
14. Agencies represented at the Forum include Solicitor General Canada and Justice Canada, the RCMP, Canada Customs and Revenue Agency, the Criminal Intelligence Service of Canada, provincial and state police, the US Border Patrol, Immigration and Naturalization Service and the US DOJ.
15. More details about the ACFTF can be found at: *www.accc.gov.au/content/index.phtml/itemId/781937/fromItemId/622554*.
16. Industry Search Australia, *Fraud is big business in Australia - Consumer be aware, 10 April 2006, www.industrysearch.com.au/news/viewrecord.aspx?id=19946*.
17. The full text of the Convention is available at: *http://conventions.coe.int/Treaty/en/Treaties/Html/185.htm*.
18. Cybercrime Convention, Articles 2, 3, 8.
19. Ibid, Article 25.
20. Ibid, Article 26.
21. In 2006, Microsoft offered a contribution to the Council of Europe to finance the Convention's implementation programme. See: *http://ec.europa.eu/justice_home/news/information_dossiers/conference_integrity/doc/Presentation_Anderson.pdf*, p.4.
22. To date, the Convention has been ratified by 18 European countries plus the United States (see: *http://conventions.coe.int* for a list).
23. EU Parliament and Council Directive 95/46 on the Protection of Individuals with Regard to the Processing of Personal Data and on the Free Movement of such Data, 24 October 1995, in *Official Journal* dated 23 November 1995, L 281, *http://eurlex.europa.eu/LexUriServ/LexUriServ.do?uri=CELEX:31995L0046:EN:HTML*.
24. Article 7(a) of EC Directive 95/46.

25. EU Parliament and Council Directive 2002/58/EC on the Processing of Personal Data and the Protection of Privacy in the Electronic Communications Sector, 12 July 2002, in *Official Journal of the European Union* ("*OJEU*"), L 201/37, at: *www.spamlaws.com/docs/2002-58-ec.pdf.*
26. See Glossary for a definition.
27. See the EC Directorate General for Information Society's webpage on *Confidentiality of Communications - spyware, cookies*, at: *http://ec.europa.eu/information_society/policy/ecomm/todays_framework/privacy_protection/spyware_cookies/index_en.htm.*
28. EU Council Framework Decision 2005/222/JHA on attacks against information systems, 24 February 2005, in *OJEU* L 69/67, at: *http://eurlex.europa.eu/LexUriServ/site/en/oj/2005/l_069/l_06920050316en00670071.pdf.*
29. Article 5 of the EC Treaty.
30. COM(2006)688final.
31. Reference to the EC FPEG may be found at: *http://ec.europa.eu/justice_home/news/information_dossiers/conference_integrity/doc/payment_fraud_en.pdf.*
32. Presentations at the conference may be found at: *http://ec.europa.eu/justice_home/news/information_dossiers/conference_integrity/interventions_en.htm.*
33. EU Parliament and Council Regulation n°2006/2004 on the Co-operation between national authorities responsible for the enforcement of consumer protection laws, 27 October 2004, in *OJEU* L364/1, at: *http://eurlex.europa.eu/LexUriServ/site/en/oj/2004/l_364/l_36420041209en00010011.pdf.*
34. Internet Changes Everything, Linkblog, *Secret Service Busts Internet Organized Crime Ring*, 29 October 2004, at: *www.ladlass.com/ice/archives/2004_10.html.*
35. The APWG maintains a website at: *www.antiphishing.org.*
36. See the APWG December 2006 report at: *www.antiphishing.org/reports/apwg_report_december_2006.pdf.*
37. DPN was established in 2004. It maintains a website at: *www.digitalphishnet.org/default.aspx.*
38. Today, access to such data is open to law enforcers, business, and the public. However, a project to restrict the access to the database is being envisaged.
39. Reference to TACD's Resolution may also be found in Section IV of this report.

40. ASC, *Best Practices: Factors for Use in the Evaluation of Potentially Unwanted Technologies*, at: *www.antispywarecoalition.org/documents/BestPractices.htm.*
41. See the website maintained by the MAAWG at: *www.maawg.org*.
42. MAAWG, *E-Marketers, Senders, ISPs Fight Spam with New MAAWG Sender Best Practices Endorsed by Industry*, at: www.maawg.org/news/maawg070515.
43. Symantec's *Online Identity Initiative* may be found at: *www.networkworld.com/news/2007/013107-demo-symantec-identity.html.*
44. See: *www.microsoft.com/presspass/press/2003/nov03/11-05AntiVirusRewardsPR.mspx.*
45. See: *http://news.com.com/Microsoft+helps+net+Bulgarian+phishers/2100-7349_3-6030016.html.*

Annex 6.1 Multilateral Instruments Addressing Online ID Theft

I. OECD instruments on E-commerce

OECD (Organisation for Economic Co-operation and Development) (1999), *Guidelines for Consumer Protection in the context of Electronic Commerce*, OECD, Paris, *www.oecd.org/document/51/0,2340,en_2649_34267_1824435_1_1_1_1,00.html.*

OECD (2003), *Guidelines for Protecting Consumers from Fraudulent and Deceptive Commercial Practices Across Borders*, OECD, Paris, *www.oecd.org/sti/consumer-policy.*

II. OECD instruments in the field of security, privacy, and spam

Security

OECD (2002), *Guidelines for the Security of Information Systems and Networks*, OECD, Paris, http://www.oecd.org/dataoecd/16/22/15582260.pdf

OECD (2007), *Recommendation and Guidance on Electronic Authentication*, OECD, Paris, *www.oecd.org/dataoecd/32/45/38921342.pdf.*

Privacy

OECD (1980), *Guidelines on the Protection of Privacy and Transborder Flows of Personal Data,* OECD, Paris, *www.oecd.org/document/18/0,3343,en_2649_34255_1815186_1_1_1_1,00.html.*

OECD (2007), *Recommendation of the Council on Cross-border Co-operation in the Enforcement of Laws Protecting Privacy*, OECD, Paris, *www.oecd.org/dataoecd/43/28/38770483.pdf.*

Spam

OECD (2006), *OECD Anti-Spam Toolkit of Recommended Policies and Measures*, OECD, Paris, *www.oecd-antispam.org/.*

III. Other international instruments

Council of Europe (2001), *Convention on Cybercrime*, Budapest, 23 November 2001, http://conventions.coe.int/Treaty/en/Treaties/Html/185.htm.

United Nation (2001), Convention against Transnational Organised Crime, 8 January 2001, www.unodc.org/pdf/crime/a_res_55/res5525e.pdf

Annex 6.2 United Nations Study on Identity Fraud

The United Nations Intergovernmental Expert Group studied fraud and criminal misuse and falsification of identity. Following are excerpts from the group's second meeting in a report to the Secretary General, 2 April 2007 (pp. 5-20):

IV. Conclusions and recommendations

B. Further work on the gathering, analysis and dissemination of information

17. The available evidence clearly suggests that economic fraud is a serious problem, and is increasing, both globally and in a number of Member States. However, many States reported that they do not have accurate information or a systematic framework for gathering and analysing such information. […] Data that would permit the quantification of fraud by occurrence or offence rates are not available in many States, almost no official data quantifying proceeds exists. […] There is growing awareness of and concern about identity-related crime, but it represents a novel concept for law enforcement and criminal justice experts in many States. There are few legislative definitions and many basic concepts remain fluid at this early stage. […]. The Intergovernmental expert group therefore made the following recommendations:

a. Further general research into economic fraud … as a global issue should be conducted […]

d. Systematic and structured processes for gathering and analysing data in each Member State are developed, and UNODC should be asked to assist in this process […] Generally, such processes should include:

(i) A standard typology or classification framework of offences or activities;

(ii) The gathering of qualitative and quantitative information from multiple sources, including official offence reports or complaints and other sources […].

(iii) To the extent feasible, the gathering and analysis of information about the costs of fraud. This would include […] the indirect economic costs, and the non-economic costs of fraud.

D. Domestic powers to investigate, prosecute and punish fraud and identity-related crime

1. Legislative measures against fraud and identity-related crime

21. […] While the vast majority of criminalisation issues appear to have been addressed, the evidence suggests that some specific enhancements could be considered to improve and modernise legislation. […]

22. Lawmakers need to develop appropriate concepts, definitions and approaches to the criminalisation of a range of conduct, including identity theft, identity fraud, and other identity-related crimes. […] It is therefore recommended that States consider the adoption of new identity-based criminal offences. It is also recommended that, in developing new offences, common approaches to criminalisation be taken, to the greatest extent possible. […]

4. Law enforcement and investigative capacity

27. […]a. It is recommended that States develop and maintain adequate research capacity to keep abreast of new developments in the use of information, communication and commercial technologies in economic fraud and identity related crime.

b. It is also recommended that the product of research be shared and disseminated throughout law enforcement in each country in domestic training, and […] with other States through appropriate technical assistance and training, and with relevant commercial entities.

d. It is further recommended that States support and make use of the "24/7" emergency contact network in transborder cybercrime matters, both for emergency and non-emergency cases involving electronic fraud or identity-related crimes.

5. Co-operation between criminal justice systems and the private sector

29. […] [It is] essential that criminal justice and commercial entities co-operate effectively, both to develop an accurate […] picture of the problems, and […] to implement preventive and reactive measures.

30. To prevent fraud and identity related crime, it is important that security countermeasures be developed and incorporated into commercial technologies and practices when they are first developed. This requires consultation between public entities […] and private interests […]

32. A key element of prevention is the education and training of persons in a position to identify and report economic fraud […], ranging from commercial customers or communication subscribers to employees […]. Such training and education requires frequent updating, to ensure that information reflects the latest developments in criminal methods and techniques, law enforcement measures, and commercial practices. It is therefore recommended that criminal justice and commercial entities co-operate […].

7. Recommendations for the prevention and deterrence of economic fraud and identity-related crime

[…] Close collaboration between relevant public and private sector entities in developing and implementing preventive measures will also be important for success […]. It is therefore recommended that Member States develop and implement effective fraud-prevention measures, at the national, regional and global levels, and in co-operation with the private sector. Such prevention efforts could include the following.

(a) The dissemination of information about fraud to potential victims […].

(b) The dissemination of information about fraud to others who may be in a position to identify, report and prevent frauds when they occur. Many commercial entities already train key employees in areas such as banking, credit card transactions […].

(c) The rapid and accurate gathering and analysis of information to support […] prevention measures. This should include the gathering of information among law enforcement, commercial and other entities at the domestic level and […] at the international level.

(d) The rapid sharing of information among appropriate law-enforcement and private sector entities at both the domestic and international levels. […].

(e) The development of commercial practices and systems […]. Effective collaboration between governments and the private sector are essential […].

35. A number of responses mentioned a range of technical means of prevention [...]. These included measures to make documents such as passports or credit cards more reliable as a means of identifying individuals and more difficult to alter or falsify. [...] The establishment of stronger identification systems in every State would bring collective benefits for the international community in controlling crimes such as economic fraud, immigration or travel-related crimes such as trafficking in persons, and in general security.

> (c). Government and commercial entities should co-operate to ensure that identification systems are robust and inter-operable to the extent that that is feasible.

36. [...] Further study and consideration of deterrence measures is therefore recommended. In addition to offences and punishments, this could include measures such as focused and specialised law enforcement units trained to deal with fraud cases, where these are seen as increasing the probability of detection, prosecution and punishment.

8. Training

37. One issue raised by some of the responses is the need for training investigators and prosecutors, and the need for technical assistance for developing countries in this area. [...] Those investigating identity-related crime require knowledge not only of crimes such as impersonation and forgery, but also knowledge of the identification infrastructures and systems which support both government and commercial forms of identification. [...]

Chapter 7. Online Identity Theft: What Can Be Done?

In recent years, a number of OECD member countries have put programmes in place for addressing ID theft (see Chapter 4). Such programmes, which tend to have strong educational and awareness aspects, target broad audiences including consumers, key employees from the public and private sector and law enforcement.

An analysis of the challenges being faced suggests that efforts to combat online ID theft have three key aspects:

Prevention – what stakeholders can do to lower the risk of identities being stolen (*e.g.* ways to enhance identity security, ways to identify attempts and instances of identity theft, and ways to limit the magnitude and scope of incidents).

Deterrence – what stakeholders can do to discourage parties from engaging in ID theft (*e.g.* legal sanctions).

Recovery and redress – what stakeholders can do to facilitate recovery and redress of such harms as financial detriment, injury to reputation, and other non-monetary harms.

This chapter summarises preventative measures for online ID theft that could be taken in two key areas: (1) education; and (2) identity authentication. The chapter concludes by identifying additional ways to combat online ID theft that would be beneficial. While these suggestions are geared to online ID theft, it should be noted that many of the measures suggested are equally applicable to offline ID theft.

Enhancing education and awareness

Educating consumers, business, government officials, and the media, and raising awareness about online ID theft are indispensable to reducing risks of theft. Reducing these risks would strengthen consumer trust in E-commerce. As stated in the *1999 OECD Guidelines*, "Governments, businesses and consumers representatives should work together to educate consumers about electronic commerce... to increase business and consumer awareness of the consumer protection framework that applies to their online activities" (OECD, 1999, Section VIII).

This recommendation, which also appears in the *2003 Guidelines* (OECD, 2003, Section II. F), is directly relevant to online ID theft. Online ID theft is a fraudulent activity which has become increasingly complex, relying on ever changing high-tech methods. Tackling it requires concerted, collaborative efforts by all stakeholders (*i.e.* government, business, and consumers). Education and awareness are therefore necessary to ensure that both consumers and businesses are aware of the importance of the problem, and knowledgeable about its evolving forms.

Structuring education and awareness programmes

Effective education and awareness programmes require: *i*) development of compelling and informative educational materials; and *ii*) development of institutions and techniques to deliver the materials and education to stakeholders in efficient ways. Moreover, co-operation and co-ordination of initiatives among parties can provide important opportunities to exploit synergies and strengthen efforts. It is thus important to involve stakeholders at an early point in developing programmes; insights from different perspectives can help to better determine what the precise education/awareness needs are, what the target audiences might be, and how they could best be reached.

Collection of relevant information on online ID theft

The collection and dissemination of basic information on online ID theft are key to raising awareness and knowledge of the importance of the problem and ways to combat it. There are five basic types of information that it would be useful to develop: *i*) statistical information showing developments and trends; *ii*) information on the non-economic effects of ID theft; *iii*) factual material on the methods that parties are using to steal identities, *iv*) general tips on how to protect identity, including tools that consumers and business could use to block online intrusions, and *v*) information on techniques that can be used to identify or recognise efforts to misuse identity information.

Statistical information showing developments and trends

In introducing and maintaining an effective framework to limit the incidence of fraudulent practices against consumers, the 2003 OECD *Guidelines* call on member countries to provide for "effective mechanisms to adequately investigate, preserve, obtain and share relevant information and evidence relating to occurrences of fraudulent ... practices" (OECD, 2003, Section II. A. 2).

Awareness of the scope and scale of the problem is a key element in support of education campaigns. However, to date, information on developments and trends in online ID theft is not generally available, despite growing member country warnings that it is on the rise. Moreover, when data are available, they rarely include sufficient detail on online forms of ID theft (OECD, 2008).

It would be beneficial for stakeholders to explore ways to enhance the development of statistical information tracking developments in ID theft. It would be helpful if this information provided specific information on online ID theft. One of the indicators that has been used in this regard is the number of consumer complaints. It would be interesting to explore what other indicators may be helpful.

In addition to measuring the magnitude of ID theft, it could be useful to monitor its economic impact on individuals and countries. Such information would further highlight and illustrate the scale of the problem.

Information that is comparable from one country to another and from different sources within one country would enhance its value. The development of such information should, where possible, draw on the efforts of multilateral groups (both public and private) that are active in the area.

Private-sector platforms could be used to gather, analyse and disseminate phishing, spam and virus statistics on a worldwide basis. These could include: the Anti-Phishing Working Group ("APWG," at *www.antiphishing.org*), which focuses on eliminating fraud and ID theft resulting from phishing and e-mail spoofing of all types; the Messaging Anti-Abuse Working Group ("MAAWG," at *www.maawg.org*), which aims at preserving the electronic messaging from online exploits and abuse such as messaging spam, malware attacks and other forms of abuse; and DigitalPhishNet ("DPN," *www.digitalphishnet.org/default.aspx*), which is a collaborative forum where Internet Service Providers, online auction sites, financial institutions, and law enforcement agencies share statistics and best practices in real time to tackle phishing and other online threats.

Information on the indirect effects of ID theft

In addition to having economic costs, ID theft can have other effects including the time victims spend to restore their reputation, the negative effects on their reputation, and the subsequent difficulties they have to re-establish creditworthiness. Collection of such information would help provide a more complete understanding of the implications of ID theft, thereby helping to raise awareness of the problems that theft can cause.

Factual material on the techniques used to steal identities

Identifying the different techniques used to commit ID theft is crucial to effectively deterring and responding to the threat. To be useful, information on these techniques needs to be collected, analysed and updated on a regular basis to keep abreast of developments. Where possible, it would be beneficial to have such information processed and shared between and among not only consumer protection enforcement actors, but also other enforcement bodies addressing the ID theft issue. ID theft indeed raises, in many cases, security, privacy and spam issues. Over the past years, ID thieves have shown a certain degree of ingenuity to get access to personal information. As indicated above, increasingly, malware and spam have been coupled with phishing.

As described in Part I of this book, so-called phishing attacks have become ever more sophisticated, taking a variety of forms and targeting both fixed and mobile electronic devices.

It should be noted that all stakeholders can play a role in developing and sharing information on the methods and techniques being employed. To maximise the utility of information that is collected, it is important that mechanisms be in place to facilitate the sharing of the information in an effective manner.

Information on the level of sophistication of online ID techniques

In addition to understanding the process by which online ID theft can be committed, education campaigns need to warn consumers about the ever-evolving forms of these methods. Phishing messages used to be quite unsophisticated and mostly text-based. For example, through the so-called "419 scam" (also well known as the online "Nigerian scam" or offline "Nigerian letter"), phishers tried to commit advance fee fraud by requesting upfront payment or money transfer from their targets. They usually offered to share a large amount of money with their potential victims that they would transfer out of their country. Victims were then asked to pay fees, charges or taxes to help release or transfer the money. However, victim of its own success, this scam is today very well known among Internet users and is not used as much anymore.

Thus, understanding the need for more complex scams, phishers have developed new ways to trick consumers into revealing passwords, bank account numbers and other personal data. Phishing scams now increasingly contain well-designed images and logos copied from legitimate commercial institutions. They have also become more personalised, sometimes containing the first digits of their targets' credit card numbers - which may actually be found on all credit cards of the same bank – to further convince their potential victim that the message is coming from their own bank. Similar to real commercial offers, phishing scams contain multiple solicitations inviting targets to reveal the password, age, address, *etc.*

And while phishers traditionally used well-known top level domain names such as ".com", ".biz" or ".info", they now attempt to avoid detection by using domain names from small island countries, such as ".im" from the UK Isle of Man, which are in many cases unknown to spam filters (McAfee, 2006, p. 15). Some phishing scams now even contain self-signed digital certificates to use the "HTTPS" security protocol and trigger the security padlock icon on spoofed websites.

Keeping consumers and other stakeholders informed of new and evolving techniques is key to enhancing prevention.

General tips on how to protect identity while on line

Providing stakeholders with practical advice on ways to protect their identities (see Box 7.1) can contribute significantly to lowering the risk of, or preventing, online ID theft. A number of organisations and governments have developed tips in these areas. One of the most comprehensive and extensive initiatives was undertaken by the United States Government, which maintains a website providing information on ways to protect personal information and avoid Internet fraud (*http://onguardonline.gov/index.html*), including ID theft.

Dissemination of information

Assuring that stakeholders are aware of, and have ready access to information on ID theft is key to enhancing prevention. At the very least, such information should be available on the Internet. In addition, orientation or training sessions in schools or on a group basis would be beneficial. Moreover, television and radio also provide opportunities to engage the public, as would the availability of printed or electronic materials (*e.g.* CD and DVD). Finally, Internet service providers and heavily frequented websites, such as search engines or auction sites, could serve as an important vehicle for pointing consumers to relevant information developed by governments and other interested parties.

Box 7.1 Consumer anti-phishing tips from OnGuardOnline.gov

- Install anti-virus and anti-spyware software, as well as a firewall on your fixed or mobile device and update them regularly.
- Avoid clicking on a link in a message that you think is spam and also make it a policy never to respond to e-mail or pop up messages that ask for your personal or financial information. Also, do not cut and paste a link from the message into your web browser. Phishers can make links look like they go one place, but then they actually take you to a look-alike site.
- Never disclose your credit card number or security digits in response to a message you suspect is spam. If you are concerned about your account, contact the organisation using a phone number you know to be genuine, or open a new Internet browser session and type in the company's correct Web address yourself.
- Forward the phishing scam to law enforcers and/or to industry groups such as the APWG, DPN or MAAWG. You may also forward phishing e-mails to the FTC at spam@uce.gov. In addition to industry groups and law enforcement agencies, you may also forward the phishing e-mail to the organisation that is being spoofed.

Co-ordination of education initiatives

Co-ordination of education and awareness initiatives provides opportunities for enhancing their effectiveness, especially to the extent that it increases coherence and simplifies efforts. Such co-ordination can take place within and between the private and government sectors, on local, national and international platforms. Such co-ordination would help identify and expand the use of particularly effective practices. Internet service providers, for example, are in an excellent position to highlight the importance of online ID theft, and point subscribers to educational resources.

It should be noted that education and awareness initiatives are multifaceted; within government, for example, the training of persons responsible for enforcement of laws covering ID theft is an important aspect of enhancing awareness and limiting the magnitude and scope of ID theft. A number of countries are already active on this front.

International law enforcement networks such as the International Consumer Protection Enforcement Network ("ICPEN") and the London Action Plan ("LAP") could be used as platforms to help co-ordinate and disseminate educational information across OECD member countries (OECD, 2003, Section III. D).

Authentication and data security

Data security is also a key component of any strategy to combat ID theft. Data compromises have many harmful consequences, including exposing consumers to the threat of ID theft, exposing the entity whose system was breached to legal liability for failure to secure the data, and imposing the risk of substantial costs for all parties involved. Accordingly, member countries should develop and ensure compliance with data security safeguards, such as laws and regulations, industry standards and guidelines, and private contractual arrangements that impose data security requirements, including, if appropriate, initiating investigations and enforcement actions against entities that violate the laws governing data security.

- Member countries should better educate the private sector on safeguarding data and encourage organisations that collect and maintain sensitive consumer information to implement practical security measures to protect consumer data.

Electronic authentication

Electronic authentication has been recognised as a useful process permitting the verification and management of identities on line. Under the 2006 OECD *Guidance on Electronic Authentication,* which sets out a number of operational principles aimed at helping member countries establish or modernise their approaches to authentication, the concept is understood as a function for establishing the validity and assurance of a claimed identity of a user, device or another entity in an information or communication system. As such, it can be an effective deterrent to the theft or misuse of personal information.

Education on the benefits and proper uses of authentication are critical for user confidence on line.

As set forth in the 2007 *OECD Recommendation on Electronic Authentication* encouraging member countries to establish compatible, technology-neutral approaches for effective domestic and cross-border electronic authentication of persons and entities, OECD countries should take steps to raise the awareness of all participants of the benefits of the use of electronic authentication at both domestic and international levels.

Electronic authentication is today considered as an element of the emerging concept of identity management. Such a broader system, which would seek to allow users to interact on line while minimising the amount of personal information they reveal on line, will be the subject of strong consideration by OECD countries in the years to come.

Areas for further work

As indicated at the outset, there are three key aspects of combating online ID theft: *i*) prevention, *ii*) deterrence and *iii*) recovery and redress. This chapter focuses on prevention, looking specifically at ways that consumers and other stakeholders could be educated to prevent online ID theft.

There is, however, a pressing need to address other aspects of the issue. Outside the OECD, the United Nations Office on Drugs and Crimes ("UNODC") is co-operating with the United Nations Commission for International Trade Law ("UNCITRAL"), developing recommendations for best practices in the prevention, deterrence, and recovery from ID theft.

The European Commission is working on a harmonised definition of the concept and is considering whether online ID theft should be criminalised throughout the EU. As indicated in Chapters 4, 5 and 6 of this book, work is also being carried out within many OECD governments by different agencies, and by the private sector.

Some of the issues that need to be addressed at the domestic and international levels (by the OECD and other international bodies) include:

- Legal
  - Should ID theft be defined legally as a specific offence?
  - What sorts of dissuasive sanctions might be appropriate (such as fines, confiscations, black lists, etc.)?
  - What legal remedies should be available for victims?
  - Should legislation require companies to take more steps to prevent identity theft, such as disclosing data security breaches affecting their customers when those breaches could result in identity theft, or improving authentication of consumers and customers when providing services or transactions?

- Cross-border enforcement co-operation between and among consumer protection enforcement authorities and the private sector.
  - How could cross-border co-operation among enforcement authorities be strengthened in the following areas?
  - Investigative and information sharing powers with foreign authorities, business and industry, consumer representatives.
  - Assistance, training, and support of other countries' law enforcement efforts.
  - Implementation and exchange of "best practices" in the area of consumer education.
- Identity recovery and redress
  - What assistance should government, industry, and/or civil society develop to help consumers restore their identity and recover from their monetary and non-monetary losses resulting from ID theft?
  - Should redress mechanisms be made available for consumers, and if so, what entities should be responsible for such redress?
  - What additional tools are needed by victims to ensure that they can restore their identity and otherwise recover fully from the identity theft?

Chapter 8. Conclusions and Recommendations

The analysis in this book suggests that stakeholders should consider addressing a number of issues to improve their efforts to combat online ID theft:

- ***Definition*** – The lack of a common definition may complicate efforts to combat ID theft in a comprehensive fashion, across borders.
- ***Legal status*** – ID theft/fraud is not an offence per se under most OECD member country laws. It is a crime in a few others. Whether ID theft should be treated as a stand-alone offence, and criminalised, needs to be considered.
- ***Co-operation with private sector*** – The private sector should actively participate in the battle. member countries could consider enacting more restrictive laws increasing penalties imposed on ID thieves; engage in outreach to the private sector and encourage entities to i) launch awareness campaigns, ii) develop industry best practices and iii) develop and implement other technological solutions to reduce the incidence of ID theft.
- ***Standards*** – Member countries should examine establishing national standards for private sector data protection requirements and impose a duty to disclose data security breaches on companies and other organisations storing data about their customers.
- ***Statistics*** – ID theft (whether offline or online) has failed to attract the attention of statisticians. Most data are from the United States; statistics for Europe do not exist, except for the United Kingdom. When data are available, they often do not cover ID theft as an independent wrong. The United States is notably one of the few countries with data available that analyse ID theft as a separate offense. The production of more tailored and accurate statistical data, covering all OECD member countries could help determine the impact of ID theft in the digital marketplace.

- ***Victim assistance*** – member countries could consider developing victim assistance programs to help victims of ID theft recover and minimise their injury.
- ***Remedies*** – Member countries could consider whether to enact legislation to provide more effective legal remedies for victims of ID theft.
- ***Deterrence and enforcement*** – The lack of criminal laws prohibiting ID theft and the limited resources of law enforcement authorities may mean that there is insufficient deterrence. Member countries could explore the value of increasing resources for law enforcement, ID theft investigations and training. More generally, given the rapid evolution of ID theft techniques and methods, more resources and training could be granted to all OECD authorities involved in the battle.
- ***Education*** – Consideration could be given to broadening education on ID theft so as to cover all interested stakeholders including consumers, users, governments, businesses, and industry.
- ***Co-ordination and co-operation*** – Agencies involved in the enforcement of anti-ID theft rules and practices are numerous at both domestic and international levels. Their respective roles and framework for co-operation could be clarified to help enhance their effectiveness.

Consideration could be given to improving domestic law enforcement co-ordination by developing national centers dedicated to the investigation of ID theft crimes. With regard to co-ordination and co-operation with foreign law enforcement authorities, member countries could explore areas of mutual interest such as: *i*) enhancing deterrence; *ii*) expanding participation in key international instruments (*e.g.* the Council of Europe Convention on Cybercrime); *iii*) improving response to requests for investigative assistance; and *iv*) otherwise strengthening co-operation with foreign partners (*e.g.* in the fields of training law enforcement).

Part III. OECD Policy Guidance on Online Identity Theft

The following are extracts from the OECD Policy Guidance on Online Identify Theft, which followed the OECD Ministerial Meeting on the Future of the Internet Economy in Seoul, Korea on 17-18 June 2008.

I. Introduction

Identity theft ("ID theft") is a longstanding problem which, as the Internet and E-commerce have developed, has expanded to include online forms. While the scope of online ID theft appears to be limited in most countries, its implications are significant as the growing risk of such theft can undermine consumer confidence in using the Internet for E-commerce. Governments have acted to fight against such fraud (both online and offline) at the domestic and international levels. The *1999 OECD Guidelines for Consumer Protection in the Context of Electronic Commerce* ("the 1999 Guidelines") and the *2003 OECD Guidelines for Protecting Consumers from Fraudulent and Deceptive Commercial Practices Across Borders* ("the 2003 Guidelines"), for example, set out principles aimed at strengthening member countries' frameworks to fight offline and online fraud. Outside the OECD, international instruments such as the Council of Europe's *Cybercrime Convention* and the United Nations' *Convention against Transnational Organised Crime* have been developed to address the issue.

The principles in the 1999 and 2003 *Guidelines* serve as a solid basis for establishing a framework to fight online ID theft and other fraud. The purpose of this paper is to describe how the principles presented in these instruments could be elaborated to strengthen and develop effective member country strategies to combat online ID theft. It explores, in particular, how education and awareness of stakeholders could be enhanced to prevent such theft. The guidance draws largely on the research and analysis contained in a *Scoping Paper on Online Identity Theft* that was considered by the Committee on Consumer Policy in 2007 (OECD, 2008).

ID theft definition, forms and methods

ID theft occurs when a party acquires, transfers, possesses, or uses personal information of a natural or legal person in an unauthorised manner, with the intent to commit, or in connection with, fraud or other crimes. Although this definition encompasses both individuals and legal entities, focus in the present guidance is limited to identity theft affecting consumers.

Traditionally, ID theft has been committed by accessing information acquired from public records, theft of personal belongings, improper use of databases, credit cards, and checking and saving accounts and misusing that information. As described in Box 1 below, off-line, unauthorised access to personal data can be carried out by various means, including dumpster diving, payment card theft, pretexting, shoulder surfing, skimming, or business record theft.

Box 1. Traditional ways to access personal data for ID theft

Dumpster diving: generally refers to the act whereby fraudsters go through bins to collect "trash" or discarded items. It is the means that identity thieves employ to obtain copies of individuals' cheques, credit card or bank statements, or other records that contain their personal information.

Pretexting: pretexters are parties who contact a financial institution or telephone company, impersonating a legitimate customer, and request that customer's account information. In other cases, the pretext is accomplished by an insider at the financial institution, or by fraudulently opening an online account in a customer's name.

Shoulder surfing: refers to the act of looking over someone's shoulder or from a nearby location as the victim enters a Personal Identification Number ("PIN") at an ATM machine.

Skimming: the capturing of personal data from the magnetic stripes on the backs of credit cards; data is then transmitted to another location where it is re-encoded onto fraudulently made credit cards.

Business record theft: refers to situations where someone steals data from a business (*e.g.* stolen computers or files) or bribes insiders to obtain the information from the business or organisation.

On line, there are principally three methods to obtain victims' personal information (see Box 2): *i*) software designed to collect personal information is secretly installed on someone's computer or device – fixed or mobile (*i.e.* malware); *ii*) deceptive e-mails or websites are used to trick persons into disclosing personal information (*i.e.* phishing – phishing e-mails are often mass-distributed via spam; they are increasingly used to install malware on the computers of recipients.); and *iii*) computers or mobile devices are hacked into or otherwise exploited to obtain personal data.

Box 2. Online methods for stealing personal information

Malware: a general term for a software code or programme inserted into an information system in order to cause harm to that system or to other systems, or to subvert them for use other than that intended by their own users. Viruses, worms, Trojan horses, backdoors, keystroke loggers, screen scrapers, rootkits, and spyware are all examples of malware (See Glossary for definitions of these terms).

Spam: commonly understood to mean unsolicited, unwanted and harmful electronic messages (OECD, 2006c) and is increasingly being viewed as a vector for malware and criminal phishing scams.

Phishing: a method that thieves use to lure unsuspecting Internet users' personal identifying information through emails and mirror-websites which look like those coming from legitimate businesses, such as financial institutions or government agencies. Typically, a phishing attack is composed of the following steps:

The phisher sends its potential victim an e-mail that appears to be from an existing company. The e-mail uses the colours, graphics, logos and wording of the company.

The potential victim reads the e-mail and provides the phisher with personal information by either responding to the e-mail or clicking on a link and providing the information via a form on a website that appears to be from the company in question.

Through this, the victim's personal information is directly transmitted to the scammer.

Hacking: exploiting vulnerabilities in electronic systems or computer software to steal personal data.

Prevalence

ID theft is an increasing problem victimising individuals across all ages and social categories. Box 3 describes the ways that identity thieves misuse consumers' personal information both off line and on line. Online ID theft has been recognised as the source of growing concerns for consumers in recent years, having a direct impact on E-commerce transactions, including mobile commerce (OECD, 2006c, p. 21). As noted in the *EU 2006 Special Eurobarometer* (European Commission, 2006, p. 12), the use of the Internet to purchase goods and services online is rather limited (only 27% of the EU population in 2005), and is mostly restricted to domestic commerce. Such limited use reflects, in part, consumers' lack of trust in E-commerce transactions, fearing that their personal information could be stolen.[1]

Box 3. Traditional and online methods of misusing personal information

Misuse of existing accounts: Identity thieves use victims' existing accounts, including credit card accounts, cheque/savings accounts, telephone accounts (both landline and wireless service), Internet payment accounts, E-mail and other Internet accounts, and medical insurance accounts.

Opening new accounts: Identity thieves use victims' personal information to open new accounts, including telephone accounts (both landline and wireless service), credit card accounts, loan accounts, cheque and savings accounts, Internet payment accounts, auto insurance accounts, and medical payment accounts.

Commit other frauds: Identity thieves also misuse victims' personal information by giving it to the police when stopped or charged with a crime, by using it to obtain medical treatment, services, or supplies, by using it in rental housing situations, by using it to obtain government benefits, and by using it in employment situations.

Efforts to combat ID theft

In recent years, a number of member countries have put programmes in place for addressing ID theft. Such programmes, which tend to have strong educational and awareness aspects, target broad audiences including consumers, key employees from the public and private sector and law enforcement. An analysis of the challenges being faced suggests that efforts to combat online ID theft have three key aspects:

Prevention – what stakeholders can do to lower the risk of identities being stolen (*e.g.* ways to enhance identity security, ways to identify attempts and instances of identity theft, and ways to limit the magnitude and scope of incidents).

Deterrence – what stakeholders can do to discourage parties from engaging in ID theft (*e.g.* legal sanctions).

Recovery and redress – what stakeholders can do to facilitate recovery and redress of such harms as financial detriment, injury to reputation, and other non-monetary harms.

This guidance focuses on the prevention of the acquisition of personal information in the online environment. Section II provides ideas on how stakeholders can use education and enhanced awareness to *i*) help consumers avoid falling victim to ID theft and *ii)* help business and government fight more effectively against the problem. Section III deals specifically with initiatives that could be taken to educate business on ways to improve data security, while Section IV addresses issues related to identity authentication. Finally, Section V identifies areas where further work on ways to combat online ID theft would be beneficial. While the guidance is geared to online ID theft, it should be noted that many of the measures suggested are equally applicable to offline ID theft.

II. Ways that education and awareness could be enhanced to prevent online ID theft

Educating consumers, business, government officials, and the media, and raising awareness about online ID theft are indispensable to reducing risks of theft. Reducing these risks would strengthen consumer trust in E-commerce. As stated in the *1999 Guidelines*, "Governments, businesses and consumers representatives should work together to educate consumers about electronic commerce... to increase business and consumer awareness of the consumer protection framework that applies to their online activities" (OECD, 1999, Section VIII). This recommendation, which also appears in the *2003 Guidelines* (OECD, 2003, Section II. F), is directly relevant to online ID theft. Online ID theft is a fraudulent activity which has become increasingly complex, relying on ever changing high-tech methods. Tackling it requires concerted, collaborative efforts by all stakeholders (*i.e.* government, business, and consumers). Education and awareness are therefore necessary to ensure that both consumers and businesses are aware of the importance of the problem, and knowledgeable about its evolving forms.

Structuring education and awareness programmes

Effective education and awareness programmes require: *i)* development of compelling and informative educational materials; and *ii)* development of institutions and techniques to deliver the materials and education to stakeholders in efficient ways. Moreover, co-operation and co-ordination of initiatives among parties can provide important opportunities to exploit synergies and strengthen efforts. It is thus important to involve stakeholders at an early point in developing programmes; insights from different perspectives can help to better determine what the precise education/awareness needs are, what the target audiences might be, and how they could best be reached.

Collection of relevant information on online ID theft

The collection and dissemination of basic information on online ID theft are key to raising awareness and knowledge of the importance of the problem and ways to combat it. There are five basic types of information that it would be useful to develop: *i*) statistical information showing developments and trends; *ii*) information on the non-economic effects of ID theft; *iii*) factual material on the methods that parties are using to steal identities, *iv*) general tips on how to protect identity, including tools that consumers and business could use to block online intrusions, and *v*) information on techniques that can be used to identify or recognise efforts to misuse identity information.

Statistical information showing developments and trends

In introducing and maintaining an effective framework to limit the incidence of fraudulent practices against consumers, the 2003 *Guidelines* call on member countries to provide for "effective mechanisms to adequately investigate, preserve, obtain and share relevant information and evidence relating to occurrences of fraudulent ... practices" (OECD, 2003, Section II. A. 2). Awareness of the scope and scale of the problem is a key element in support of education campaigns. However, to date, information on developments and trends in online ID theft is not generally available, despite growing member country warnings that it is on the rise. Moreover, when data are available, they rarely include sufficient detail on online forms of ID theft (OECD, 2008).

It would be beneficial for stakeholders to explore ways to enhance the development of statistical information tracking developments in ID theft. It would be helpful if this information provided specific information on online ID theft. One of the indicators that has been used in this regard is the number of consumer complaints. It would be interesting to explore what other indicators may be helpful.

In addition to measuring the magnitude of ID theft, it could be useful to monitor its economic impact on individuals and countries. Such information would further highlight and illustrate the scale of the problem.

Information that is comparable from one country to another and from different sources within one country would enhance its value. The development of such information should, where possible, draw on the efforts of multilateral groups (both public and private) that are active in the area. Private-sector platforms could be used to gather, analyse and disseminate phishing, spam and virus statistics on a worldwide basis. These could include: the Anti-Phishing Working Group ("APWG," at *www.antiphishing.org*), which focuses on eliminating fraud and ID theft resulting from phishing and e-mail spoofing of all types; the Messaging Anti-Abuse Working Group ("MAAWG," at *www.maawg.org*), which aims at preserving the electronic messaging from online exploits and abuse such as messaging spam, malware attacks and other forms of abuse; and DigitalPhishNet ("DPN," *www.digitalphishnet.org/default.aspx*), which is a collaborative forum where Internet Service Providers, online auction sites, financial institutions, and law enforcement agencies share statistics and best practices in real time to tackle phishing and other online threats.

Information on the indirect effects of ID theft

In addition to having economic costs, ID theft can have other effects including the time victims spend to restore their reputation, the negative effects on their reputation, and the subsequent difficulties they have to re-establish creditworthiness. Collection of such information would help provide a more complete understanding of the implications of ID theft, thereby helping to raise awareness of the problems that theft can cause.

Factual material on the methods and techniques that parties are using to steal identities

Identifying the different techniques used to commit ID theft is crucial to effectively deterring and responding to the threat. To be useful, information on these techniques needs to be collected, analysed and updated on a regular basis to keep abreast of developments. Where possible, it would be beneficial to have such information processed and shared between and among not only consumer protection enforcement actors, but also other enforcement bodies addressing the ID theft issue. ID theft indeed raises, in many cases, security, privacy and spam issues. Over the past years, ID thieves have shown a certain degree of ingenuity to get access to personal information. As indicated above, increasingly, malware and spam have been coupled with phishing.

As described in Box 4 below, phishing attacks have become ever more sophisticated, taking a variety of forms and targeting both fixed and mobile electronic devices.

It should be noted that all stakeholders can play a role in developing and sharing information on the methods and techniques being employed. To maximise the utility of information that is collected, it is important that mechanisms be in place to facilitate the sharing of the information in an effective manner.

Box 4. Phishing variants

Pharming: this method, which uses the same kind of spoofed identifiers as in a classic phishing attack, redirects users from an authentic website (e.g. a bank website) to a fraudulent site that replicates the original. When the customer connects its computer to its bank web server, a hostname lookup is performed to translate the bank's domain name (e.g. "bank.com") into an IP address. During that process, the IP address will be changed.

SmiShing: cell phone users receive text messages ("SMS") where a company confirms their signing up for one of its dating services and that they will be charged a certain amount per day unless they cancel their order at the company's website. Such a website is in fact compromised and used to steal personal information.

Vishing: in a classic spoofed e-mail, appearing from legitimate businesses or institutions, the phisher invites the recipient to call a telephone number. When calling, the target reaches an automated attendant, requesting personal data such as account number, or password for pretended "security verification" purposes. Victims feel usually safer in this way as they are not required to go to a website to transmit their personal information.

Information on the level of sophistication of online ID techniques

In addition to understanding the process by which online ID theft can be committed, education campaigns need to warn consumers about the ever evolving forms of these methods. Phishing messages used to be quite unsophisticated and mostly text-based. For example, through the so-called "419 scam" (also well known as the online "Nigerian scam" or offline "Nigerian letter"), phishers tried to commit advance fee fraud by requesting upfront payment or money transfer from their targets. They usually offered to share a large amount of money with their potential victims that they would transfer out of their country. Victims were then asked to pay fees, charges or taxes to help release or transfer the money. However, victim of its own success, this scam is today very well known among Internet users and is not used as much anymore.

Thus, understanding the need for more complex scams, phishers have developed new ways to trick consumers into revealing passwords, bank account numbers and other personal data. Phishing scams now increasingly contain well-designed images and logos copied from legitimate commercial institutions. They have also become more personalised, sometimes containing the first digits of their targets' credit card numbers - which may actually be found on all credit cards of the same bank – to further convince their potential victim that the message is coming from their own bank. Similar to real commercial offers, phishing scams contain multiple solicitations inviting targets to reveal the password, age, address, *etc.*

And while phishers traditionally used well-known top level domain names such as ".com", ".biz" or ".info", they now attempt to avoid detection by using domain names from small island countries, such as ".im" from the UK Isle of Man, which are in many cases unknown to spam filters (McAfee, 2006, p. 15). Some phishing scams now even contain self-signed digital certificates to use the "HTTPS" security protocol and trigger the security padlock icon on spoofed websites.

Keeping consumers and other stakeholders informed of new and evolving techniques is key to enhancing prevention.

General tips on how to protect identity while on line

Providing stakeholders with practical advice on ways to protect their identities (see Box 5) can contribute significantly to lowering the risk of, or preventing, online ID theft. A number of organisations and governments have developed tips in these areas. One of the most comprehensive and extensive initiatives was undertaken by the United States Government, which maintains a website providing information on ways to protect personal information and avoid Internet fraud (*http://onguardonline.gov/index.html*), including ID theft.

Box 5. Consumer anti-phishing tips from OnGuardOnline.gov

Install anti-virus and anti-spyware software, as well as a firewall on your fixed or mobile device and update them regularly.

Avoid clicking on a link in a message that you think is spam and also make it a policy never to respond to e-mail or pop up messages that ask for your personal or financial information. Also, do not cut and paste a link from the message into your web browser. Phishers can make links look like they go one place, but then they actually take you to a look-alike site.

Never disclose your credit card number or security digits in response to a message you suspect is spam. If you are concerned about your account, contact the organisation using a phone number you know to be genuine, or open a new Internet browser session and type in the company's correct Web address yourself.

Forward the phishing scam to law enforcers and/or to industry groups such as the APWG, DPN or MAAWG. You may also forward phishing e-mails to the FTC at spam@uce.gov. In addition to industry groups and law enforcement agencies, you may also forward the phishing e-mail to the organisation that is being spoofed.

Dissemination of information

Assuring that stakeholders are aware of, and have ready access to information on ID theft is key to enhancing prevention. At the very least, such information should be available on the Internet. In addition, orientation or training sessions in schools or on a group basis would be beneficial. Moreover, television and radio also provide opportunities to engage the public, as would the availability of printed or electronic materials (*e.g.* CD and DVD). Finally, Internet service providers and heavily frequented websites, such as search engines or auction sites, could serve as an important vehicle for pointing consumers to relevant information developed by governments and other interested parties.

Co-ordination of education and awareness initiatives

Co-ordination of education and awareness initiatives provides opportunities for enhancing their effectiveness, especially to the extent that it increases coherence and simplifies efforts. Such co-ordination can take place within and between the private and government sectors, on local, national and international platforms. Such co-ordination would help identify and expand the use of particularly effective practices. Internet service providers, for example, are in an excellent position to highlight the importance of online ID theft, and point subscribers to educational resources.

It should be noted that education and awareness initiatives are multifaceted; within government, for example, the training of persons responsible for enforcement of laws covering ID theft is an important aspect of enhancing awareness and limiting the magnitude and scope of ID theft. A number of countries are already active on this front.

International law enforcement networks such as the International Consumer Protection Enforcement Network ("ICPEN") and the London Action Plan ("LAP") could be used as platforms to help co-ordinate and disseminate educational information across OECD member countries (OECD, 2003, Section III. D).

III. Data security

Data security is also a key component of any strategy to combat ID theft. Data compromises have many harmful consequences, including exposing consumers to the threat of ID theft, exposing the entity whose system was breached to legal liability for failure to secure the data, and imposing the risk of substantial costs for all parties involved. Accordingly, member countries should develop and ensure compliance with data security safeguards, such as laws and regulations, industry standards and guidelines, and private contractual arrangements that impose data security requirements, including, if appropriate, initiating investigations and enforcement actions against entities that violate the laws governing data security.

- Member countries should better educate the private sector on safeguarding data and encourage organisations that collect and maintain sensitive consumer information to implement practical security measures to protect consumer data.

IV. Electronic authentication

Electronic authentication has been recognised as a useful process permitting the verification and management of identities on line. Under the 2006 OECD *Guidance on Electronic Authentication,* which sets out a number of operational principles aimed at helping member countries establish or modernise their approaches to authentication, the concept is understood as a function for establishing the validity and assurance of a claimed identity of a user, device or another entity in an information or communication system. As such, it can be an effective deterrent to the theft or misuse of personal information.

Education on the benefits and proper uses of authentication are critical for user confidence on line.

As set forth in the 2007 *OECD Recommendation on Electronic Authentication* encouraging member countries to establish compatible, technology-neutral approaches for effective domestic and cross-border electronic authentication of persons and entities, OECD countries should take steps to raise the awareness of all participants of the benefits of the use of electronic authentication at both domestic and international levels.

Electronic authentication is today considered as an element of the emerging concept of identity management. Such a broader system, which would seek to allow users to interact on line while minimising the amount of personal information they reveal on line, will be the subject of strong consideration by OECD countries in the years to come.

V. Further work

As indicated at the outset, there are three key aspects of combating online ID theft: *i)* prevention, *ii)* deterrence and *iii)* recovery and redress. This paper focused on prevention, looking specifically at ways that consumers and other stakeholders could be educated to prevent online ID theft. There is, however, a pressing need to address other aspects of the issue. Outside the OECD, the United Nations Office on Drugs and Crimes ("UNODC") is co-operating with the United Nations Commission for International Trade Law ("UNCITRAL"), developing recommendations for best practices in the prevention, deterrence, and recovery from ID theft. The European Commission is working on a harmonised definition of the concept and is considering whether online ID theft should be criminalised throughout the EU. As indicated in the *Scoping Paper on Online Identity Theft* (OECD, 2008), work is also being carried out within many OECD governments by different agencies, and by the private sector.

Some of the issues that need to be addressed at the domestic and international levels (by the OECD and other international bodies) include:

- Legal:
  - Should ID theft be defined legally as a specific offence?
  - What sorts of dissuasive sanctions might be appropriate (such as fines, confiscations, black lists, etc.)?
  - What legal remedies should be available for victims?
  - Should legislation require companies to take more steps to prevent identity theft, such as disclosing data security breaches affecting their customers when those breaches could result in identity theft, or improving authentication of consumers and customers when providing services or transactions?
- Cross-border enforcement co-operation between and among consumer protection enforcement authorities and the private sector.
  - How could cross-border co-operation among enforcement authorities be strengthened in the following areas?
  - Investigative and information sharing powers with foreign authorities, business and industry, consumer representatives.
  - Assistance, training, and support of other countries' law enforcement efforts.
  - Implementation and exchange of "best practices" in the area of consumer education.

- Identity recovery and redress
  - What assistance should government, industry, and/or civil society develop to help consumers restore their identity and recover from their monetary and non-monetary losses resulting from ID theft?
  - Should redress mechanisms be made available for consumers, and if so, what entities should be responsible for such redress?
  - What additional tools are needed by victims to ensure that they can restore their identity and otherwise recover fully from the identity theft?

Note

1. A 2006 *International Telecommunication Union Online Survey* (ITU, 2006) concluded that more than 40% of Internet users refrain from transacting on line for that reason.

Glossary

Backdoors: A malicious software program that allows an attacker to access a system by listening to commands on a certain User Diagram Protocol ("UDP") or Transmission Control Protocol (TCP) port. Backdoors facilitate the attacker's acquisition of information such as passwords and allows the attacker to execute remote commands.

Bots and botnets: Some malware is distributed using botnets, a group of "zombies" or bots infected computers compromised though malware and turned into malware that can be used to carry out attacks against other computer systems. These computers become compromised when a bot program, a type of malware, is installed on the system.

Dumpster diving: Generally refers to the act whereby fraudsters go through bins to collect "trash" or discarded items. It is the means that identity thieves employ to obtain copies of individuals' cheques, credit card or bank statements, or other records that hold their personal information.

Keystroke loggers: A program that records and "logs" how a keyboard is used. There are two types of keystroke loggers. The first type of keystroke logger requires the attacker to retrieve the logged data from the compromised system. The second type of keystroke logger actively transmits the logged data.

Man-in-the-middle attack: The process by which the phisher collects personal data through the interception of an Internet user's message that was intended to be sent to a legitimate site.

Pharming: The use of deceptive e-mail messages to redirect users from an authentic website to a fraudulent one, which replicates the original in appearance.

Phishing: The use deceptive e-mails to get users to divulge personal information, includes luring them to fake bank and credit-cards websites.

Pretexting: A form of social engineering used to obtain sensitive information. In many instances, pretexters contact a financial institution or telephone company, impersonating a legitimate customer, and request that customer's account information. In other cases, the pretext is accomplished by an insider at the financial institution, or by fraudulently opening an online account in the customer's name.

Rootkit: A set of programs designed to conceal the compromise of a computer at the most privileged base or 'root' level. As with most malware, rootkits require administrative access to run effectively, and once achieved can be virtually impossible to detect.

Shoulder surfing: In relation to ID theft, refers to the act of looking over someone's shoulder or from a nearby location as the victim enters her Personal Identification Number ("PIN") at an ATM machine.

Skimming: The recording of personal data from the magnetic stripes on the backs of a credit cards; data is then transmitted to another location where it is re-encoded onto fraudulently made credit cards.

SMiShing: The sending of text messages ("SMS") to cell phone users that trick them into going to a website operated by the thieves.

Spam: Commonly understood to mean unsolicited, unwanted and harmful electronic messages. here appears to be a growing correlation between malware and spam.

Spyware: A form of malware that sends information from a computer to a third party without the user's permission or knowledge. Different types of Spyware may collect different types of information. Some Spyware tracks the websites a user visits and then sends this information to an advertising agency while malicious variants attempt to intercept passwords or credit card numbers as a user enters them into a web form or other applications.

Trojan horse: A computer program that appears legitimate but which actually has hidden functionality used to circumvent security measures and carry out attacks. Typically a Trojan enters a user's computer by exploiting a browser vulnerability or feature.

Virus: A hidden software program that spreads by infecting another program and inserting a copy of itself into that program. A virus requires a host program to run before the virus can become active. The term "virus" is increasingly used more generically to refer to both viruses and worms.

Vishing: Phishing via Voice over Internet Protocol ("VoIP").

VoIP: A new technique using phones to steal individuals' personal information.

Worm: A type of malware that self replicates without the need for a host program. Worms can exploit weaknesses in a computer's operating system or other installed software and spread rapidly via the Internet. A mass-mailing worm is a worm that is spread out through bulk e-mail.

Bibliography

ACCC (Australian Competition and Consumer Commission) (2003), Court declares imitation Sydney Opera House website illegal, press release, 28 August 2003: *www.accc.gov.au/content/index.phtml/itemId/360431/fromItemId/378016* .

ACPR (Australian Centre for Policing Research) (2006), Review of the legal status and rights of victims of identity theft in Australasia, Report Series No. 145.2, Commonwealth of Australia: *www.acpr.gov.au/pdf/ACPR145_2.pdf.*

ANSI (American National Standards Institute) and BBB (Better Business Bureau) (2008) ANSI-BBB Identity Theft Prevention and Identity Management Standards Panel Final Report, 31 January 2008, *www.ansi.org/standards_activities/standards_boards_panels/idsp/report_webinar08.aspx?menuid=3.*

APEC (2006), Letter of Support from the Chair of the Telecommunications and Information Working Group, 20 March 2006, Strasbourg: *www.coe.int/T/E/Legal_affairs/Legal_cooperation/Combating_economic_crime/6_Cybercrime/T-CY/.*

APEC (Asian-Pacific Economic Co-operation) (2005), Strategy to Ensure a Trusted, Secure and Sustainable Online Environment, November 2005: *www.apec.org/apec/apec_groups/working_groups/telecommunications_and_information.html.*

APWG (Anti-Phishing Working Group) (2006a), Phishing Activity Trends, report for November 2006: *www.antiphishing.org/reports/apwg_report_november_2006.pdf.*

APWG (2006b), Phishing Activity Trends, report for December 2006: *www.antiphishing.org/reports/apwg_report_december_2006.pdf.*

APWG (2007), Phishing Activity Trends, report for April 2007: *www.antiphishing.org/reports/apwg_report_april_2007.pdf.*

ASWPRPCC (Australasian and South West Pacific Region Police Commissioners' Conference) (2005), Australasian Identity Crime Policing Strategy 2006-2008, report produced by the ACPR, December: 2005: *www.acpr.gov.au/pdf/ID%20Crime%20Strat%2006-08.pdf.*

British Telecom, CPP, Get Safe Online, Lloyds TSB, Metropolitan Police, Yahoo! (UK) (2006): Security Report, February 2006, *www.btplc.com/onlineidtheft/onlineidtheft.pdf.*

BWGCBMMF (2004), Report on Identity Theft, report to the Ministry of Public Safety and Emergency Preparedness Canada and the Attorney General of the United States, October 2004, *www.ps-sp.gc.ca/prg/le/bs/report-en.asp.*

BWGCBMMF (Bi-national Working Group on Cross-Border Mass Marketing Fraud) (US-Canada) (2006), Report on Phishing, October 2006, report to the Ministry of Public Safety and Emergency Preparedness Canada and the Attorney General of the United States: *www.psepcsppcc.gc.ca/prg/le/_fl/Phishing%20for%20CBCF%202006-en.pdf.*

CAO (Cabinet Office) (Japan) (2006), Summary Report on the Enforcement Status of Act on the Protection of Personal Information in Fiscal Year 2005, June 2006: *www5.cao.go.jp/seikatsu/kojin/foreign/enforcement-status2005.pdf.*

CMC (Consumer Measures Committee) (Canada) (2005), Working Together to Prevent Identity Theft, A discussion paper for public consultation, 6 July 2005: *http://cmcweb.ic.gc.ca/epic/site/cmc-cmc.nsf/vwapj/Discussion%20Paper_IDTheft.pdf/$FILE/Discussion%20Paper_IDTheft.pdf.*

Deloitte Touche Tohmatsu, 2006 Global Security Survey: *www.deloitte.com/dtt/cda/doc/content//CA_FSI_2006%20Global%20Security%20Survey_2006-06-13.pdf.*

ENISA (European Network and Information Security Agency) (2006), Survey on Industry Measures taken to comply with National Measures implementing Provisions of the Regulatory Framework for Electronic Communications relating to the Security of Services, 2006: *www.enisa.europa.eu/pages/05_01.htm.*

European Commission (2007), Communication from the Commission to the European Parliament, the Council and the Committee of the Regions, Towards a General Policy on the Fight against Cyber Crime, 22 May 2007, COM(2007) 267 FINAL, *http://eurlex.europa.eu/LexUriServ/site/en/com/2007/com2007_0267en01.pdf.*

European Commission (2006), DG SANCO, Special Eurobarometer, Consumer Protection in the Internal Market, September 2006, Brussels, *http://ec.europa.eu/public_opinion/archives/ebs/ebs252_en.pdf.*

European Commission (2004), Identity Theft: A Discussion Paper, Joint Research Centre, Institute of the Protection and Security of the Citizen, EUR 21098 EN, 2004.

European Commission FPEG (EC Fraud Prevention Expert Group) (2007), Report on Identity Theft/Fraud, FPEG, subgroup on identity theft, 22 October 2007: *http://ec.europa.eu/internal_market/fpeg/docs/id-theft-report_en.pdf.*

Europol (2006), EU 2006 Organised Crime Threat Assessment ("OCTA"): *www.europol.eu.int/publications/OCTA/OCTA2006.pdf.*

GetSafeOnline (UK) (2006), The Get Safe Online Report, October 2006: *www.getsafeonline.org/media/GSO_Cyber_Report_2006.pdf.*

Home Office Identity Fraud Steering Committee (UK) (2006), Identity Crime Definitions: *www.identitytheft.org.uk/definition.html.*

IDTTF (Identity Theft Task Force) (US) (2007), Combating Identity Theft: A Strategic Plan, 23 April 2007: *www.idtheft.gov.*

INTERVICT (International Victimology Institute Tilburg) (2006), The Challenge of Countering Identity Theft, Report Commissioned by the National Infrastructure Cyber Crime program ("NICC"), 6 September 2006, *www.tilburguniversity.nl/intervict/publications/NicolevanderMeulen.pdf*

ITRC (Identity Theft Resource Center) (US) (2004), Identity Theft: the Aftermath 2004, September 2005: *www.idtheftcenter.org/prteen1006.pdf.*

ITTC (Identity Theft Technology Council) (US), Online Identity Theft: Phishing Technology, Chokepoints, and Countermeasures, 3 October 2005, *www.antiphishing.org/Phishing-dhs-report.pdf.*

International Telecommunication Union (2006), Cybersecurity Awareness Survey, results as of 17 May 2006, *www.itu.int/newsroom/wtd/2006/survey/charts/q_8.asp.*

Javelin Strategy and Research (2006), 2006 Identity Fraud Survey Report, *www.javelinstrategy.com/products/AD35BA/27/delivery.pdf.*

Javelin Strategy and Research (2007), 2007 Identity Fraud Survey Report - Identity Fraud Is Dropping, Continued Vigilance Necessary, Consumer Version, February 2007, *www.javelinstrategy.com/uploads/701.R_2007IdentityFraudSurveyReport_Brochure.pdf.*

McAfee Avert Labs (2004), Anti-Phishing: Best Practices for Institutions and Consumers, White Paper, September 2004, *www.antiphishing.org/sponsors_technical_papers/AntiPhishing_Best_Practices_for_Institutions_Consumer0904.pdf.*

McAfee Avert Labs (2007), Identity Theft, White Paper, January 2007, *www.mcafee.com/us/threat_center/white_paper.html.*

McAfee (2006), Virtual Criminality Report, December 2006, *www.sigma.com.pl/pliki/albums/userpics/10007/Virtual_Criminology_Report_2006.pdf.*

Microsoft (2006), presentation by Nancy Andersen, Microsoft Vice-President, contribution to the European Commission's conference on "Maintaining the integrity of identities and payments: Two challenges for fraud prevention," The Threat of Cybercrime: The Challenge of Online Identity Theft and Strengthening the Public-Private Partnership in a Changing Threat Environment, 22 November 2006, Brussels, *http://ec.europa.eu/justice_home/news/information_dossiers/conference_integrity/doc/Presentation_Anderson.pdf*

NCL (National Consumer League) (UK) (2006), A Call for Action: Report from the National Consumer League, Anti-Phishing Retreat, Washington D.C., March 2006, *www.nclnet.org/news/2006/Final%20NCL%20Phishing%20Report.pdf.*

OECD (2006c), OECD Anti-Spam Toolkit of Recommended Policies and Measures, OECD, Paris, *www.oecd-antispam.org/.*

OECD (2009) Computer Virues and Other Malicious Software: A Threat to the Internet Economy.

OECD (2007d), The Development of Policies for the Protection of Critical Information Infrastructures, OECD, Paris, [DSTI/ICCP/REG(2007)20/FINAL].

OECD (1999), Guidelines for Consumer Protection in the context of Electronic Commerce, OECD, Paris, *www.oecd.org/document/51/0,2340,en_2649_34267_1824435_1_1_1_1,00.html.*

OECD (2003), Guidelines for Protecting Consumers from Fraudulent and Deceptive Commercial Practices Across Borders, OECD, Paris: *www.oecd.org/sti/consumer-policy.*

OECD (1980), Guidelines on the Protection of Privacy and Transborder Flows of Personal Data, OECD, Paris: *www.oecd.org/document/18/0,3343,en_2649_34255_1815186_1_1_1_1,00.html.*

OECD (2002), Guidelines for the Security of Information Systems and Networks, OECD.

OECD (2006b), Mobile Commerce, DSTI/CP(2006)7/FINAL, Directorate for Science, Technology and Industry, *www.oecd.org/sti/consumer-policy.*

OECD (2006b), OECD Anti-Spam Toolkit of Recommended Policies and Measures, OECD, Paris: *www.oecd-antispam.org/.*

OECD (2006d), Protecting Consumers from Cyberfraud, OECD Policy Brief, Paris, October 2006: www.oecd.org/sti/crossborderfraud.

OECD (2007c), Recommendation and Guidance on Electronic Authentication, OECD, Paris: *www.oecd.org/dataoecd/32/45/38921342.pdf.*

OECD (2007b), Recommendation of the Council on Cross-border Co-operation in the Enforcement of Laws Protecting Privacy, OECD, Paris: *www.oecd.org/dataoecd/43/28/38770483.pdf.*

OECD (2007), Recommendation on Consumer Dispute Resolution and Redress, OECD, Paris, *www.oecd.org/dataoecd/43/50/38960101.pdf.*

OECD (2006a), Report on the Implementation of the 2003 OECD Guidelines for Protecting Consumers from Fraudulent and Deceptive Commercial Practices Across Borders, OECD, Paris, www.oecd.org/dataoecd/45/53/37125909.pdf.

OECD (2008), Scoping Paper on Online Identity Theft, DSTI/CP(2007)3/FINAL, Directorate for Science, Technology and Industry.

OECD PUBLISHING, 2, rue André-Pascal, 75775 PARIS CEDEX 16
PRINTED IN FRANCE
(93 2009 02 1 P) ISBN 978-92-94-05658-9 – No. 56605 2009